Technical Communication for
Environmental Action

SUNY series, Studies in Technical Communication

Miles A. Kimball and Charles H. Sides, editors

Technical Communication for Environmental Action

Edited by

Sean D. Williams

Published by State University of New York Press, Albany

For information, contact State University of New York Press, Albany, NY
www.sunypress.edu

Library of Congress Cataloging-in-Publication Data

Name: Williams, Sean D., editor
Title: Technical communication for environmental action
Description: Albany : State University of New York Press, [2023] | Series:
  SUNY series, studies in technical communication | Includes bibliographical
  references and index.
Identifiers: ISBN 9781438491295 (hardcover : alk. paper) | ISBN 9781438491288
  (pbk. : alk. paper) | ISBN 9781438491301 (ebook)
Further information is available at the Library of Congress.

10 9 8 7 6 5 4 3 2 1

For Heidi who stands behind everything.
For Brendan and Nathan
who live their ideals today for a better tomorrow.

Contents

Illustrations

Figures

Tables

Acknowledgments

I would like to acknowledge the National Consortium of Environmental Rhetoric and Writing (NCERW) for sponsoring me as writer-in-residence as I prepared the manuscript for this volume.

Time spent alone in the desert does wonders to enliven the mind.

Introduction

SEAN D. WILLIAMS

Climate change is one of the most significant challenges facing the global community in the twenty-first century, and with its position at the border between people, technology, science, and communication, technical communication has a significant role to play in helping to solve these complex environmental problems. Curiously, however, technical communication (TC) research has remained relatively quiet on the ways our field contributes to positive environmental action. To help invigorate the conversation in TC about environmental issues, this collection of essays amplifies the work of scholars engaging with these challenges by creating a conversation about the ways that our field has contributed to pragmatic and democratic action to address climate change. Tillery (2018) makes a compelling argument that the history of TC scholarship reveals a tradition of mostly theoretical and rhetorical analysis of environmental communication (cf. Killingsworth & Palmer, 1992; Herndl & Brown, 1996; Ross, 2017). In comparison to that tradition, this collection explores the actual *practice* of technical communicators participating in community projects, government processes, nonprofit programs, and international work that shapes environmental action. Because of its focus on action, this collection addresses the need identified by Simmons (2008) in *Participation and Power*, and by Coppola and Karis (2000) in *Technical Communication, Deliberative Rhetoric, and Environmental Discourse* to examine firsthand cases of technical communication driving environmental action.

Beyond simple descriptions of the work we do, therefore, the collection foregrounds practical applications of technical communication theories such as social justice, participatory design, community action, service learning, and ethics to construct a *praxis* of environmental action in TC. Importantly, this collection locates *praxis* within *phronesis*—practical wisdom of communities—to consider the local and cultural values that must inform effective and ethical environmental action. Finally, *dialogue* lies at the core of this collection since only good-natured collaboration can begin to help solve problems as complex as climate degradation. By focusing the collection on these three concepts—praxis, phronesis, and dialogue—the collection hopes, as the title suggests, to paint a picture of the ways that technical communicators participate in shaping environmental action.

As a field, technical communication has been usefully described as a profession in which information is made accessible and usable for those who need that information to accomplish their goals—we are user advocates who filter, architect, and design information solutions to complex problems (*Defining Technical Communication*, n.d.). Complementing this practical orientation, TC retains a significant humanistic and civic orientation that Carolyn Miller (1989) wonderfully described many years ago as "a matter of conduct rather than of production, as a matter of arguing in a prudent way toward the good of the community rather than of constructing texts." TC, therefore, is a pragmatic discipline that seeks to drive concrete action by engaging with communities to define and describe problems, to invent solutions, and ultimately to hold one another accountable for the success of the interventions we collaborated to create.

Technical communication praxis, therefore, is the first key concept that establishes a unique scope for this collection. Commonly understood as putting theories or concepts into actual practice (Miller, 1989; Sullivan et al., 1997; Moore & Richards, 2018), praxis might more usefully be described as theoretically informed action (Katz, 1992). As a field, TC has a substantial body of theories, including, for example, participatory design (cf. Spinuzzi, 2005); user-centered design (cf. Zachry & Spyridakis, 2016); and iterative design (cf. Mayhew 1999) on one hand, and socio-cultural concerns such as social justice (cf. Jones et al., 2016); community engagement (cf. Simmons 2008); and ethics (Dombrowski, 2000) on the other hand. By focusing on praxis, this collection emphasizes theoretically informed action occurring in the real world with real people.

Understanding the values of real people and how they contribute to environmental action represents a second key concept appearing in the chapters of this collection. Compared to generalizable scientific and economic theories that often are used in environmental decision making, phronesis draws on the knowledge and wisdom of specific groups in specific locations (Kinsella & Pitman, 2012). Including the authentic, lived experiences of actual people who might be affected by a particular environmental challenge echoes TC's emphasis on participatory design and foregrounds the complex negotiations that must occur between diverse groups vying, first, to define problems and, second, to explore solutions to those problems. While some excellent prior scholarship discusses the importance of democratic participation in environmental action (cf. Grabill & Simmons, 1998; Blythe et al., 2008; Simmons 2008), the chapters in this collection take an additional step by explicitly foregrounding the ways that technical communicators rely on the practical wisdom of a polyphony of voices—phronesis—to create ethical and effective communication products that drive positive environmental action.

Finally, *dialogue* complements the prior two concepts of praxis and phronesis as the theoretical underpinning of this collection because it carries the ideas of both "practical" and "contextual." As a practice, dialogue requires that people make meaning in specific contexts, with specific words, with specific perspectives, and foregrounds the idea of building shared understandings (Grice, 1989; Sperber & Wilson, 1986). These shared understandings rely on people working in good faith to build common vocabularies, references, frames, and values by placing their individual understandings into dialogue with another as they build new understandings (Habermas, 1984). Unfortunately, much scholarship on environmental communication, both in TC and in journalism/mass communication, originates in the "deficit model" where scientists or experts know best, and the communicator's job is to translate that knowledge to others. Instead, dialogue teaches us that no "deficit" exists, just different conceptualizations of similar issues. Competing stakeholders don't share the same vocabulary or values and so simply "translating the science" inevitably will lead to misunderstandings because the model is presumed to be one-way where the nonexperts "receive" the information (cf. Perrault, 2013). By comparison, dialogue requires that parties openly engage in conversation based on principles of respect and equity: people need to talk *with* one another—not *at* one another—to build understanding. This collection intentionally foregrounds the practice of dialogue to help craft ethical and effective environmental action.

In short, while retaining careful grounding in technical communication theories, this collection explores the actual practice of technical communicators participating in shaping environmental action. Doing so helps us extend the role our profession plays not only in translating science and technology to the public—a traditional role of TC—but also in promulgating the important democratic values of participatory design and environmental justice both in conceptualizing complex environmental challenges and in crafting prudent, pragmatic solutions to those challenges that genuinely respect the dialogue among the multiple forms of knowledge and expertise held by diverse stakeholders: indigenous populations, environmentalists, individuals, businesses, governments, nonprofits, landowners, and many others.

Perhaps now, more than ever in history, technical communicators have a responsibility to employ our unique skills at connecting people, technology, science, and communication to drive global environmental action. This collection is one response to this obligation.

Interacting with This Collection

Because this collection emphasizes dialogue, readers have many possible ways to explore the excellent work of the contributors. Dialogue doesn't proceed in a linear way in the natural world—it follows its own path depending upon those interacting and how they seek to build a shared perspective. This collection intentionally follows this model. Some readers might choose to read this collection from beginning to end—and that would be perfectly fine—because the chapters are ordered to foreground a conversation among two, three, or four authors on a particular type of *agency* (Wilson, 2001; Stephens & DeLorme, 2019). Agency, of course, intersects closely with *social justice* (Jones et al., 2016; Colton & Holmes, 2018; Sackey, 2018) and given this collection's focus on action and empowering people to act on behalf of the environment, every chapter in this collection takes up these intersecting themes of agency and social justice in one way or another.

However, the richness of the contributions to this collection enables other possible pathways through the book where readers might explore chapters according to their interests. The alternative reading pathways presented in table I.1 and described later in this introduction provide some possible ways to interact with this collection—a sort of academic "Choose Your Own Adventure"—that might respond to many readers' unique concerns about the intersection of technical communication and environmental action. A summary of each chapter follows the pathway descriptions so that readers have yet another opportunity to construct their own pathway through the collection according to their individual interests if they choose not to engage with one of the four alternative possibilities sketched in table I.1.

Table I.1. Alternative Reading Pathways for the Collection

Diverse Voices	Chapter 1	Chapter 2	Chapter 3	Chapter 6	Chapter 7	Chapter 10
Narrative Methods	Chapter 1	Chapter 4	Chapter 5	Chapter 6	Chapter 8	Chapter 11
Policy & Process	Chapter 2	Chapter 3	Chapter 4	Chapter 7	Chapter 9	Chapter 10
Pedagogy	Chapter 3	Chapter 5	Chapter 8	Chapter 11		

Diverse Voices

This group of chapters draws on recent calls for technical communication to be more inclusive (cf. Agboka, 2021; Gonzales, 2021; Itchuaqiyaq & Matheson, 2021). Specifically, these chapters ask readers to consider perspectives that are absent—or mostly absent—from scholarship in TC, to include viewpoints from outside North America and Western Europe, the traditional foci of most literature in the field. These chapters also include marginalized voices from within North America ranging from Inuit peoples to midwestern farmers.

Narrative Methods

Research methods in technical communication have expanded in recent years with many scholars adopting narrative methods to complement more traditional qualitative and quantitative forms of research (cf. Jones, 2017; Williams et al., 2016). This group of chapters demonstrates the power of narratives for understanding and representing TC knowledge, and in many ways complicates accepted forms of knowing in TC by challenging readers to ask the question, "What counts as research and knowledge?" for a topic as complex as environmental action.

Policy and Process

The environment is often regarded as a resource to be exploited for the benefit and comfort of human beings, and that perspective means governments, corporations, and diverse communities (among others) must negotiate how to manage the natural world (Killingsworth & Palmer, 1992; Herndl & Brown, 1996; Simmons, 2008; Ross, 2017). As this group of chapters demonstrates, those negotiations often take the form of debate about regulatory oversight, including who is empowered to participate in processes for constructing policies and the consequences that those policies and processes have for the environment, including nonhuman species.

Pedagogy

For technical communication to positively impact the environment, teachers must inspire others to act and demonstrate that what occurs in the classroom can have real consequences in the world. TC has a strong history of service learning, client-based learning, and community engagement (Matthews & Zimmerman, 1999; Bowdon & Scott, 2003; Youngblood & Mackiewicz, 2013), and this group of chapters presents cases derived from classroom practice that not only demonstrate the significant contribution students can make but also offer practical advice on integrating environmental action into the TC classroom.

Chapter Summaries

In the essay that opens this collection, "When the Sound Is Frozen: Extracting Climate Data from Inuit Narratives," Cana Uluak Itchuaqiyaq speaks about bodily knowledges as the basis for understanding the natural world: when your breath freezes, for example, the temperature is –50 Fahrenheit or when ice breaks under the weight of a snowmobile, two adults, and four seals, the ice measures between three and four inches thick. We tell stories—narratives—about the natural world that draw on our lived, embodied experiences, and those experiences connect to traditional knowledge that exists within a community. Yet, Itchuaqiyaq argues, "science" often disregards the bodily knowledge presented through stories: "Narratives are a method of communicating important local expertise that is often overlooked as 'scientific' knowledge," they argue. Importantly, the chapter itself is a narrative about the author's experiences, about how "we Inuit have been taught to value and develop our relationship with the land, the waters, and the ice that surround us. We have been taught to listen to its stories." Yet that intimate knowledge of the environment seems to matter little in decision making when "experts" visit the Arctic with their ideas about what is best for the Inuit community. Instead, Itchuaqiyaq argues, outsiders should ask the community what it needs, what it values, what it knows—outsiders should listen—because those who live on this land know better than anyone how best to protect it.

Dan Card's chapter that follows, "Boundary Waters: Deliberative Experience Design for Environmental Decision Making," echoes Itchuaqiyaq's concern about whose views matter for environmental action. In this chapter, Card traces the processes currently unfolding (in 2021) that will determine the fate of the pristine Boundary Waters region of Minnesota and whether it will be opened to new mining activities. Drawing on principles of participatory design, Card argues that "the importance of local, context-specific praxis . . . begins with understanding local geographies, local communities, and local impacts. In short, designing ethical and effective decision-making processes requires we listen and work to understand the unique local dimensions of a given environmental problem." Card asks, do environmental review processes engage the right people in dialogue at the right time? He investigates the complexities of the review process for the Twin Metals project, notably contrasting the actual review process with an ideal one, building a framework to ensure environmental justice. Technical communication for environmental action, in Card's view, requires careful attention to process because equitable processes ensure we are working with the right people and asking the right questions to generate positive outcomes for the natural world.

Bob Hyland recounts how he has used his courses to inspire students to generate positive environmental outcomes in his chapter, "In Defense of a Greenspace: Students Discover Agency in the Practice of Community-Engaged Technical Communication." The essay begins by connecting service learning, student identity formation, and advocacy in technical communication literature, then turns to discuss agency and why we must help students move from passive critics resigned that climate change "just is" to agents of change "bending the arc of justice." Moving to a close analysis of student work produced over a series of three semesters to protect an "urban oasis" in Cincinnati, Hyland demonstrates that when students see real outcomes of their community-engaged classwork, they develop a strong sense of agency, and that technical communication has a responsibility to undertake environmental action by engaging students: "If we frame for our students that TC *can* be used to make a difference, we may be missing an opportunity for the educement of agency. Instead, I'm positing that we provide opportunities for our students to see that TC *must* be used to make a difference." Our pedagogies must demonstrate real impacts, not hypothetical ones, in other words. Educing agency, Hyland continues, requires localizing TC pedagogy and coursework because showing how we can positively impact our local communities empowers students to think about global challenges. If students recognize that technical communication products such as infographics and technical reports possess the power to influence a community to preserve a small, wooded area; to sustain a community partnership; to restore the ecology of the greenspace; then perhaps those students will feel empowered to undertake larger challenges, to employ the practice of technical communication for global environmental action.

Daniel Richards's essay, "Flood Insurance Rate Maps as Communicative Sites of Pragmatic Environmental Action," introduces the idea that technical communicators can undertake "procedural, banal work . . . as we figure out the bigger, more wicked things." The chapter narrates the author's experiences leading a project to construct an online tool that sought to persuade residents of a coastal community prone to flooding to purchase flood insurance policies. This tool, a literal calculator to help residents determine their yearly insurance premiums, represents a core technical communication practice: translating the abstruse appendix J of the National Flood Insurance Program's (NFIP) rate tables into plain language usable by nonexperts to make informed decisions about how to act. Richards demonstrates, however, that the apparent simplicity of the task hid extraordinary complexity for him as a design team of one—a role familiar to many technical communicators, especially those working in contract or freelance positions similar to the government-funded project that Richards describes. The essay traces the author's learning path about concepts like "occupancy type," "structure type," and "BFE-base flood elevation," all concepts he had to understand to write accessible documentation for nonexpert users to review as they priced insurance. Writing was the easy part, though; coding all the complex decision trees proved to be beyond the author's capabilities, and as the project reached its deadline, he had to enlist the help of a programmer to make the website work as he imagined. Richards concludes that despite the difficulties he faced creating the flood rate calculator, he learned an important lesson: "technical communication cracks the objectivist sheen of regulatory writing. Projects highlighting these realities and advocating for change are in great need." Demystifying the regulatory complexity of flood insurance stands as a metaphor for many forms of environmental action because this apparently banal communication helps communities become resilient, recognizing the need to protect people from the harm of advancing flood waters.

"Collaborating for Clean Air: Virtue Ethics and the Cultivation of Transformational Service-Learning Partnerships," by Lauren Cagle and Roberta Burnes, undertakes the challenging work of theorizing a "participatory action teaching" partnership. In their autoethnography of an emerging partnership between a university professor and an environmental educator working for the state of Kentucky, the authors take a unique look not at what it takes for students to collaborate with external partners, but at the emerging relationship of those who facilitate the student learning. In other words, this essay describes how the authors' collaboration evolved from a "normal" service-learning project into something much more sophisticated, participatory action-oriented teaching "where the emphasis is on teaching, but the teaching does not happen without the participatory action. . . . It's not just about teaching within the confines of a college class, but about a messy ongoing relationship." The character of that relationship—based in the virtues of honesty, generosity, respect, humility, and justice as well as feminist community engagement—challenges the "service learning as charity model" by presenting an alternative view where the needs of students, the faculty member, the university, the community partner, and the partner's organization have equal voice in co-constructing what students do in a course. The essay's structure itself reveals this commitment to respectful dialogue with the authors weaving short, reflective narratives throughout their argument describing the concept of a transformational relationship. The authors conclude, "There's nothing uniquely environmental about this notion of participatory action teaching. And our partnership is environmental by default." However, this essay perhaps reveals a deeper logic, one we might call "ecosystem thinking," where differences coexist harmoniously, multiplying the contribution and virtue of any single participant. Importing that logic into our classrooms is itself a form of environmental action that can open "paths for scholars to prioritize relationships among humans and nonhumans alongside, or even above data," the authors suggest, because environmental action requires us to think deeply about how co-constructive, mutually respectful relationships can change the way we view the natural world and act within it.

Beth Shirley adopts a similar ecosystem logic in her essay, "The Narrative of Silent Stakeholders: Reframing Local Environmental Communications to Include Global Human Impacts," arguing for a perspective called "societal teleconnections" that examines "the faraway effects of decisions made at home." From this perspective, environmental challenges don't unfold in just one location but result from the relationships among distant places, and we must examine shared experiences of local communities and distant ones to break down the local versus global binary. To demonstrate this perspective, the author presents research conducted with a women's agricultural association in rural Morocco, describing technical communication practices within the association used both to maintain the organization and to improve the community. For example, one member of the association was able to petition the government for olive trees, empowering the women in the community to become more resilient since the trees provided habitat for bees raised by the women, for soil stabilization at their home gardens, and for shade to cool the soil. The local narratives collected by the research project, though, become most significant for environmental action when they are included, for example, in policy reports or decision-making conversations about local actions and local priorities. As Shirley argues, "Technical communicators are in a unique position to include these narratives of stakeholders who otherwise have few other ways of making their voices heard. Because we have that capability, we must make use of it when engaging environmental issues." Changing the nature of environmental debates, that is, requires us to connect "decisions made in the United States to impacts on people in underresourced communities," especially those in the Global South who are likely to be the most impacted by climate change. Through this essay, then, Shirley models the idea of societal teleconnections as it might be used in environmental action, introducing readers of this collection (who likely are reading in English and probably reside in North America) to a group of mostly silenced women in Morocco, giving voice to the environmental challenges they face and asking us to investigate our role—half a world away—for creating those problems.

Agriculture, although in the United States, also occupies a central place in "Resilient Farmland: The Role of Technical Communicators" by Sara Parks and Lee Tesdell. The essay describes possibilities for technical communication in the context of the $1.1 trillion agriculture industry in the United States and presents cases about how technical communication could better intervene to protect midwestern agricultural land by informing decision makers "about the ravages of climate change and agricultural innovations that might mitigate them." Like most essays in this collection, Parks and Tesdell argue persuasively about the importance of knowing a place to understand how environmental communication works in a particular context, specifically the row-crop farms of Iowa. The authors establish this connection to place by describing some typical genres of technical communication used in midwestern agriculture such as the "field day," which are "site visits that invite farmers, landowners, and other key people such as managers, consultants, and agents to the (usually literal) field in order to see systems, tools, and new techniques in action." Importantly, Parks and Tesdell argue, these genres are used differently than in other contexts, so technical communicators can't rely solely on their training. Instead, they must know the context and position themselves to make a difference when the *kairos* is right. The essay draws on this background about Iowa farming and common genres to describe two cases, one failed and one successful. The unsuccessful case focuses on a lawsuit brought by the Des Moines Water Works Board against three counties in Northwest Iowa based on the claim that those counties were polluting the rivers with excess nitrates from farm fields. The public communication around the lawsuit relied on technical communication products such as watershed maps, graphs of pollutants, technical descriptions of testing procedures, and websites to make the case. The successful case describes STRIPS, Science-Based Trials of Rowcrops Integrated with Prairie Strips, especially the persuasion strategies that rely on engaging stakeholders' values to promote the project. We learn from these cases and the background that leads to them that technical communication can play an important role in "turning the ship of Big Ag and Big Petroleum in the direction of policies and practices that bring us cleaner water and healthier soil," and how important this type of environmental action is because "declining soil and water quality is literally an existential threat to humankind."

Monika Smith, in the essay "Writing for Clients, Writing for Change: Proposals, Persuasion, and Problem Solving in the Technical Writing Classroom," continues the discussion about technical communication's role in changing institutional inertia, describing a TC classroom as a site of significant environmental action. Smith argues, for example, that "grounding students' grasp of resource consumption in the daily life of their own campus can shift the classroom dynamic from *getting* students to write for environmental action to *inspiring* them to do so." To establish the grounds for the argument about student agency, the author carefully links environmental action to social justice issues, then transitions to a discussion about community-engaged learning (CEL) that forms a pillar of the strategic plan at the author's university. Smith concludes, "My first tip is that it is well worth looking to your own campus for suitable and meaningful partnerships." Smith recounts how looking to her own campus generated rich opportunities for computer science students in a required technical communication course to apply their problem solving and research skills to addressing environmental challenges on their own campus such as reducing electricity consumption and reducing the amount of plastic waste going to landfills. Writing, Smith argues, for one's own campus inspired students to engage with real audiences because if their ideas were to make a difference "right here on their very own campus . . . those ideas had to resonate with the entire community: not just their campus client, but everyone who would be directly impacted by the changes they envisioned." Working on their campus, that is, enabled students to recognize that technical communication can be used for environmental action not someplace far away, but where they lived and worked every day. Writing for change and participating in concrete environmental action with a place-based ethos, Smith concludes, might help in "paving the way to a transformational project that foregrounds the environment in any decision-making process," especially those of engineers who might be disengaged from conversations about public welfare.

Public welfare is precisely the topic of Barbara George's chapter, "Health in the Shale Fields: Technical Communication and Environmental Health Risks." In this essay, the author outlines a detailed relationship between environmental justice and procedural justice that enables the author "to critique patterns of powerful interests exploiting a marginalized locality for particular resources that often result in long-term degradation of local land and water." George's critique is directed specifically at communication surrounding the risks of high-volume hydraulic fracturing—fracking—to the low-income residents of southwestern Pennsylvania who "live with the histories of extractive economies and poverty and currently lack the agency to speak about environmental justice." George compares case studies from the region employing stasis theory to demonstrate a distinction between what "is" communicated and what "ought" to be communicated. Describing communication from the Pennsylvania Department of Environmental Protection and the Pennsylvania Department of Health, the author details confusion among the agencies about responsibility for protecting the public and how these governmental agencies—supposedly charged with protecting the public—squelched debate and engagement by the very people they were meant to serve. Additionally, these agencies significantly underreported health risks about fracking, stating, for example, that more studies are required due to "limited evidence of relationships of living near [a fracking site] and poor infant health." George contrasts the governmental response to a local nonprofit that effectively—and ethically—employed technical communication to reveal the significant health risks associated with fracking. The organization, called the Southwest Pennsylvania Environmental Health Project, enables environmental justice by providing platforms for comparing the scientific literature on risks associated with fracking, connecting individuals to healthcare providers, and providing legal resources for those who believe they have been impacted by fracking. In short, the nonprofit develops technical communication from an "ought" perspective about how the local community "ought" to be treated and what resources and information "ought" to be available, providing a model for how governmental agencies might develop their communication platforms to protect both the local environment and the health of those who live there.

Josephine Walwema provides a complementary case of responsible and ethical government action to protect the environment and the community in her essay, "Participatory Policy: Enacting Technical Communication for a Shared Water Future." The chapter analyzes technical communication produced by the City of Cape Town, South Africa, two years after that city nearly ran dry, exploring how Cape Town developed a participatory approach to developing policy for "achieving a shared water future." Specifically, the author argues, "building trusts necessitates a dialogic interaction that assures the public *learns* the science and how it affects their lives, and the technical experts *control* the messaging with accurate science." That dialogue requires that the technical experts interact with the public to solicit their experiential knowledge, and this experiential knowledge in turn allows the technical experts to adapt their messaging to local contexts. This dialogic approach, Walwema demonstrates, enables the public to translate technical knowledge into action that can help regulate environmental risks in the lives of individual communities. The author uses critical discourse analysis to evaluate policy and strategic documents produce by Cape Town about its water strategy to reveal how the structure, discourse, and genres of the strategy documents reveal a dialogue between experts and the public: "The policy documents . . . are visibly informed by scientific methods, data, and measurements, but they are rendered as technical communication that deliberates on the choices offered by the science and the collective values of the people of Cape Town." Importantly, this dialogic approach provides practical guidance for environmental action as the local communities can see how their experiential knowledge intersects with technical knowledge, because "ideas are not expressed as hypothetical but as provisions . . . that predispose Cape Town residents toward practical engagement in securing a water future for their city."

Echoing Walwema's optimistic findings, Michelle Hall Kells challenges readers to drive environmental action through "the heart lines of appreciation and imagination" in her essay, "Rhino Crash: Teaching Science, Medical, and Environmental Writing for Social Action." Importantly, Kells foregrounds "interspecies kinship ties as a critical feature of environmental education in technical communication," closing a loop that returns us to Itchuaqiyaq's essay that opened the collection. The essay begins very personally, talking about Pilar, the author's wild mustang rescue, and how a serendipitous conversation with Pilar's veterinarian launched a cascade of possibilities for connecting the author's students to conservation activities for black rhinos in South Africa. Kells outlines a 10-part pedagogical praxis revolving around the concept of *ubuntu*, "an ecology of braided attachments," that views all life as inextricably bound together and structures the pedagogy around stories of hope rather than despair. The author positions each stage of the course within student work to demonstrate each component of an approach designed to inspire her class to undertake environmental action. Beyond creating flyers, posters, and online products to stir awareness about the plight of rhinos, the students generated a grant proposal for an art installation and created a student club sanctioned by the author's university dedicated to protecting rhinos. Kells argues, "We learned that rhinos, as endangered species, index a larger ecosystem in distress wherein threatened human and nonhuman species together face habitat deterioration, natural resource depletion, poaching, predation, displacement, and mutual extinction." The essay concludes, however, on the optimistic tone where it began, showing that if Pilar, the author's mustang, could recover from what was expected to be a fatal set of afflictions, then perhaps transformation—"restoring our planet and all our relations"—through environmental action is possible and that "biophilia, the love of life in all its forms, is the quintessential response to our collective suffering."

Finally, in her epilogue, "Right Relation with the Whole World: Creating a Richer Polyvocality for Environmental Technical Communication," Caroline Gottschalk Druschke reflects on the themes and contributions of this collection. Her reflection lands on four key themes—innovative process, scalar connection, improvised action, and right relation—that weave the chapters together and concurrently serve as "a manifesto for the future of environmental technical communication: offering a roadmap for where the field has been, is now, and might—or even *must*—be headed in response to the increasingly urgent demands of environmental degradation and conflict." We must, the author argues, find ways for technical communication to place humans into "right relation" with the whole world.

As you read these chapters—in whatever order seems most appropriate—I hope that you will be inspired to undertake environmental action in some form. Technical communication's intersecting concern of people, technology, science, and communication demands that we act—*that we do something*—to make the world a better place, not just for humans, but for all life on the planet. The environmental actions we take now could save hundreds or thousands of threatened species of plants and animals across the globe from extinction, of which we humans are just one. Humans have incredible creative potential, and we should direct our energy to conserving, not consuming; to aiding, not harming; to offering kindness as a first response. The essays in this collection offer some points to begin conversations about technical communication's responsibility for our collective well-being and show the tight connection between agency and social justice and how environmental action can participate in both. This collection marks only the beginning of a conversation that *must* evolve into concrete action for the environment.

How will you participate?

References

Agboka, G. Y. (2021). "Subjects" in and of research: Decolonizing oppressive rhetorical practices in technical communication research. *Journal of Technical Writing and Communication, 51*(2), 159–174.

Blythe, S., Grabill, J. T., & Riley, K. (2008). Action research and wicked environmental problems: Exploring appropriate roles for researchers in professional communication. *Journal of Business and Technical Communication, 22*(3), 272–298.

Bowdon, M. A., & Scott, J. B. (2003). *Service-learning in technical and professional communication*. Longman.

Colton, J. S., & Holmes, S. (2018). A social justice theory of active equality for technical communication. *Journal of Technical Writing and Communication, 48*(1), 4–30.

Coppola, N. W., & Karis, B. (Eds.). (2000). *Technical communication, deliberative rhetoric, and environmental discourse: Connections and directions*. Greenwood.

Defining Technical Communication. (n.d.) Retrieved from Society for Technical Communication. https://www.stc.org/about-stc/defining-technical-communication/

Dombrowski, P. (2000). *Ethics in technical communication*. Allyn and Bacon series in Technical Communication. Allyn and Bacon.

Gonzales, L. (2021). (Re)framing multilingual technical communication with Indigenous language interpreters and translators. *Technical Communication Quarterly, 31*(1), 1–16.

Grabill, J. T., & Simmons, W. M. (1998). Toward a critical rhetoric of risk communication: Producing citizens and the role of technical communicators. *Technical Communication Quarterly, 7*(4), 415–441.

Grice, P. (1989). *Studies in the way of words*. Harvard University Press.

Habermas, J. (1984). *The theory of communicative action* (vol. 1). Beacon Press.

Herndl, C. G., & Brown, S. C. (Eds.). (1996). *Green culture: Environmental rhetoric in contemporary America*. University of Wisconsin Press.

Itchuaqiyaq, C. U., & Matheson, B. (2021). Decolonizing decoloniality: Considering the (mis) use of decolonial frameworks in TPC scholarship. *Communication Design Quarterly Review, 9*(1), 20–31.

Jones, N. N. (2017). Rhetorical narratives of black entrepreneurs: The business of race, agency, and cultural empowerment. *Journal of Business and Technical Communication, 31*(3), 319–349.

Jones, N. N., Moore, K. R., & Walton, R. (2016). Disrupting the past to disrupt the future: An antenarrative of technical communication. *Technical Communication Quarterly, 25*(4), 211–229.

Katz, S. B. (1992). The ethic of expediency: Classical rhetoric, technology, and the Holocaust. *College English, 54*(3), 255–275.

Killingsworth, M. J., & Palmer, J. S. (1992). *EcoSpeak: Rhetoric and environmental politics in America*. SIU Press.

Kinsella, E. A., & Pitman, A. (Eds.). (2012). *Phronesis as professional knowledge: Practical wisdom in the professions* (vol. 1). Springer Science & Business Media.

Matthews, C., & Zimmerman, B. B. (1999). Integrating service learning and technical communication: Benefits and challenges. *Technical Communication Quarterly, 8*(4), 383–404.

Mayhew, Deborah. (1999). *The usability engineering lifecycle: A practitioner's handbook for user interface design*. Morgan Kaufman.

Miller, C. R. (1989). What's practical about technical writing. In B. E. Fearing & W. Keats Sparrow (Eds.), *Technical writing: Theory and practice* (pp. 14–24). Modern Language Association.

Moore, K. R., & Richards, D. P. (Eds.). (2018). *Posthuman praxis in technical communication*. Routledge.

Perrault, S. (2013). *Communicating popular science: From deficit to democracy*. Springer.

Ross, D. (2017). *Topic-driven environmental rhetoric*. Routledge.

Sackey, D. J. (2018). An environmental justice paradigm for technical communication. In A. Hass & M. Eble (Eds.), *Key theoretical frameworks: Teaching technical communication in the twenty-first century* (pp. 138–160). Utah State University Press/University Press of Colorado.

Simmons, W. M. (2008). *Participation and power: Civic discourse in environmental policy decisions*. State University of New York Press.

Sperber, D., & Wilson, D. (1986). *Relevance: Communication and cognition*. Harvard University Press.

Spinuzzi, C. (2005). The methodology of participatory design. *Technical Communication, 52*(2), 163–174.

Stephens, S. H., & DeLorme, D. E. (2019). A framework for user agency during development of interactive risk visualization tools. *Technical Communication Quarterly, 28*(4), 391–406.

Sullivan, P., Porter, J. E., & James Porter, S. (1997). *Opening spaces: Writing technologies and critical research practices*. Greenwood.

Tillery, D. (2018). *Commonplaces of scientific rhetoric in environmental discourses*. Routledge.

Williams, S. D., Ammetller, G., Rodríguez-Ardura, I., & Li, X. (2016). A narrative perspective on international entrepreneurship: Comparing stories from the United States, Spain, and China. *IEEE Transactions on Professional Communication, 59*(4), 379–397.

Wilson, G. (2001). Technical communication and late capitalism: Considering a postmodern technical communication pedagogy. *Journal of Business and Technical Communication, 15*(1), 72–99.

Youngblood, S. A., & Mackiewicz, J. (2013). Lessons in service learning: Developing the service-learning opportunities in technical communication (SLOT-C) database. *Technical Communication Quarterly, 22*(3), 260–283.

Zachry, M., & Spyridakis, J. H. (2016). Human-centered design and the field of technical communication. *Journal of Technical Writing and Communication, 46*(4), 392–401.

Chapter 1

When the Sound Is Frozen

Extracting Climate Data from Inuit Narratives

Cana Uluak Itchuaqiyaq

Reader, I am writing to you, colleague to colleague, human to human, heart to heart, to tell you a story about stories. The way I tell this story may bend the rules you are used to when discussing "academic" things because this story is beyond that level of abstraction. This is a human story and whether you recognize it yet or not, you are a character in its plot.

This story goes back at least three generations in my Inuit family, and it will extend out into every generation forward. It is a story about bodies, and it is a story about ice.

~

I was born in my homelands in Northwest Alaska in January, or *Siqiñaas-rugruk*, which means the first appearance of the sun. In January in the Alaskan Arctic, the sun remains low on the horizon, and it is cold. My great-uncle, Qiġñak, described how the Ipani, the Iñupiat from long ago, would tell one another about the weather: "The Ipani . . . had no way to determine the temperature, but he could tell how cold it was. If he had been out hunting and someone asked him when he came in, 'How cold is it today?' he would answer: 'I can hear my breath freezing in the cold

19

air like something sizzling. It must then be 50 to 60 below zero when you can hear your breath freezing'" (Wells, 1974, p. 11). For those who live in the Arctic today, understanding weather relies upon both reports from the National Weather Service and the ways the body perceives and encounters weather, much like my ancestor describes. For me growing up in the Arctic, I had my own ways of interpreting the concept of 50 or 60 degrees below zero based on how it felt to my body. When I was a young woman in the 1990s in the Arctic, I was too vain as well as too lazy to put on winter gear every time I had to go outside. However, when I left home wearing only jeans rather than wearing snow pants as an outer layer, I knew it was dangerously cold when it quickly felt like my skin would split if I slapped my thigh. When I felt that sensation, I knew it was 50 or 60 below and wearing only jeans outside for an extended period was not a good choice. In those moments, I used my body as an instrument to understand weather and evaluate action.

Reader, at this point I want to take a moment to recognize that you might not have experience existing in –50°F temperatures. That observation—"I do not have experience in –50°F temperatures"—provides important context for my overall argument: Inuit lived experience in Arctic environments is an essential component to scientific understanding of the Arctic. Although Indigenous knowledges and lived experiences have increasingly been incorporated into climate change research, their use has generally been relegated to only providing local context rather than scientific data (Ellam Yua et al., 2022; Simonee et al., 2021). Technical communication as a field is uniquely positioned to assist researchers in both recognizing and utilizing the rich sources of data contained in narratives describing lived experiences and observations.

In this chapter, I will describe the use of *context-based knowledges* to ascribe and extract "scientific" meaning and data from narrative utterances. What I mean by context-based knowledges is that these knowledges are subjective and specific to particular contexts. These contexts can be shared among groups of individuals with similar experiences in much the same way that an inside joke operates where a key word or phrase is used to recall a particular set of experiences to a particular group of people. These contexts can also be used to share or communicate meaning to others outside of these experiences, although it generally requires some explanation much in the same way outsiders can be let "in" on an inside joke. By comparison, *place-based knowledge* is rooted in place as a fundamental starting or connection point. Place-based knowledge assumes that individuals with experience in a place share, to some degree, similar experiences and observations of that place. Context-based knowledges, on the other hand, account for the nuances related to an individual's or group's particular experiences related to place or other such *contexts*, such as Qiġñak hearing his breath freezing in −50°F temperatures or me feeling like my thigh would split from the cold.

While that observation "I do not have experience in −50°F temperatures" indicates one point of expertise of Inuit or others living in the Arctic, it does not exclude anyone from making similar observations. Most individuals have some experience gauging temperature or weather conditions using their own contextual and bodily knowledges as the basis of such observations rather than relying on the use of thermometers or other instruments. For example, people with curly hair might understand air humidity levels based on how their hair reacts. Or perhaps you live in a location where the weather gets cold and dry. Imagine being outside in the fall and your nose starts running. That observation, "my nose started running," is easy to make because you will likely be compelled to react to your nose running by removing that mucus, say by wiping it with the side of your pointer finger and then wiping that finger on your pants. In those collective experiences, you might come to an understanding that "it is cold and dry" because "my nose started running." According to Dr. Andrew Lane of Johns Hopkins University, our nose's primary purpose is to condition the air we breathe. Noses run as a response to cold, dry air because our lungs prefer warm, moist air (NPR, 2009). Hence, you can quantify the "my nose started running" experience into a specific range of temperatures and weather data by consulting a weather app on your phone after wiping your snot onto your pants. Once you have connected a weather condition, say ~40°F and ~40% humidity, to the experience of "my nose started running" through multiple runny noses and temperature checks (see table 1.2), you can then equate "my nose started running" to "it's ~40°F and ~40% humidity."

While this might be a bit of a hard transition, I want to present two tables of possible data from what I've shared thus far to demonstrate how one might use narratives as technical communication from which to extract data. Table 1.1 presents the narrative discussing −50°F to −60°F weather as air temperature data, and table 1.2 presents an iterative approach to understanding the phenomenon "my nose started running" as contextual weather data. Granted, there are other factors that might contribute to the phenomenon "my nose started running," such as physical activity or health.

Table 1.1. Equating –50°F to –60°F to Bodily Experiences

Narrative Reference	Air Temperature	Notes
"I can hear my breath freezing in the cold air like something sizzling. It must then be 50 to 60 below zero when you can hear your breath freezing" (Wells, 1974, p. 11).	–50°F to –60°F	Qiġñak equates the sound of one's breath freezing to –50°F to –60°F.
When I . . . went outside wearing only jeans rather than wearing snow pants as an outer layer, I knew it was dangerously cold when it quickly felt like my skin would split if I slapped my thigh. When I felt that sensation, I knew it was 50 or 60 below and wearing only jeans outside for an extended period of time was not a good choice.	–50°F to –60°F	Itchuaqiyaq equates the feeling of their skin potentially splitting open at –50°F to –60°F.

If it helps conceptually, for table 1.2 imagine that the field observations are taken while sitting and watching an event outside. The point is that one can become attuned to bodily responses to external conditions and with some experience, individuals can use those responses to make more concrete observations about those conditions.

Table 1.2. Sample Field Test of Conditions Related to "My Nose Started Running"

Field Test	Conditions	Notes
"My nose started running," observation #1.	43°F, 41% humidity	Checked weather app once my nose began running for each test.
"My nose started running," observation #2.	37°F, 39% humidity	
"My nose started running," observation #3.	40°F, 40% humidity	
Results:	40°F, 40% humidity	The average results from three field tests.

My mother, I'yiiqpak, has always told me that members of our family were known for being especially attuned to the weather before technologies like thermometers or weather stations came to the Arctic. My great-grandmother, Matulik, born before white missionaries came to our lands and brought with them concepts like numbering years or assigning numbers to the surrounding air, would read the sky in the evening and the early morning and predict the weather. She could forecast storms by reading the sky, which is an important skill still used today in Inuit communities (Simonee et al., 2021). My mom remembers being a child and hearing her mother, Uluak, say, "*Matulgum taiñña siḷa tautuviksaulgitchchaa. Siḷagiiḷiñiaqpalulgitchchuq*" (Matulik is intently looking at the sky. Maybe we're coming to a bad weather.) Not only did Matulik watch the sky to predict bad weather, others in the community watched Matulik watching the sky to make their own weather predictions and to prepare accordingly. Similarly, my father, Lumen, taught me about how hunters used *aqargiq*, willow ptarmigan, to predict snowstorms. *Aqargiq* are easy to butcher in the field. You can remove their crop, a little holding sac attached to their throat, and observe its contents. If the crop is filled with food, that means that there is a storm coming because the *aqargiq* stored away food in preparation to bed down for a few days. My father would bring home these little sacs filled with small seeds and blow them up like a balloon to dry, calling them rattles, and use them to tell us a storm was coming.

Growing up in a subsistence family in the Arctic, I experienced descriptions of weather in both qualitative and quantitative ways. Sometimes these observations were presented in parallel with one another. For example, if I were to ask my mother about the weather, she might say "it's 30 below" and that the coldness of the air "just takes your breath away." Even when I do not explicitly ask her about the weather, I will often receive references to the weather as openers to her stories because it is a typical and relevant detail that is included in everyday conversation for people who live in the Arctic. In our language, Iñupiatun, we do not have a word for hello. Instead, when people greet one another in traditional ways, it is with an observation of the weather. Elders, when greeting one another in Iñupiatun, might say something along these lines: "*Anuqḷiqsinniġaa*" (It is windy). "*Ii, napmuŋnaġuminñaitpaluktuq*" (Yes, it's not a good day to go anywhere). For example, if I were to ask my mother about what she's been up to and she responds, "It is so cold. I went to the post office and the air took my breath away just walking outside that little bit," I am receiving data, or rather technical communication about weather, derived from her body and lived experiences as instruments to gauge and report on weather conditions. In this exchange, because of previous parallel observations she's made to me and through actively listening for cues about the weather, I am able to interpret the data she's presenting through narrative. Table 1.3 demonstrates the technical communication embedded within I'yiiqpak's narratives using a prior parallel observation as the basis for subsequent references to "the air takes your breath away."

The activity of the weather is both observable in the way one's body responds and in physical manifestations on the land, water, and ice. My family's home in Kotzebue, Alaska, is located on a long peninsula surrounded by the waters of the Kotzebue Sound that are visible from our home's windows. In deciding whether and when to go out in our boats on the water, my brother, Qaluraq, will look out the window and read the water before consulting the weather forecast. If there are white caps visible on waves in particular locations, he knows that it is windy and the conditions to cross the sound from our peninsula to the adjoining mainland shorelines will likely be perilous. When the sound is frozen, darkness on the horizon might indicate open leads in the ice with exposed water, or may indicate overflow, called *qaaptit*, a layer of water on top of ice, and unstable ice conditions are likely coming soon. It is common for members of rural Inuit communities to combine weather information from reading the sky, lands, and waters with weather forecasts from apps in order to make decisions about traveling out in the country because the information from the apps themselves are not reliable enough (Simonee et al., 2021).

Table 1.3. I'yiiqpak Says, "The Air Takes Your Breath Away"

Narrative Reference	Air Temperature	Notes
Parallel Observation A: "It's 30 below" and that the coldness of the air "just takes your breath away."	–30°F	When I'yiiqpak mentions that the air takes one's breath away, she equates that bodily experience to –30°F.
Parallel Observation B: "It is so cold. I went to the post office and the air took my breath away just walking outside that little bit."	–30°F	Uses Parallel Observation A that equates "the air takes your breath away" with –30°F.

I have been out in unstable ice conditions exactly once in my lifetime and that was enough for me. My role in our family is more of the resident nerd, writer, and spreadsheet-maker rather than any kind of subsistence provider. However, I once filmed a documentary that followed the harvesting of seals in my family, from ice to table, for a school project. My brother, Qaluraq, took me far out onto the ice where the seals, weighing upward of 200 pounds each, were hanging out and enjoying sunny, springtime conditions. My inexperience seal hunting is both traditional—men typically fulfilled the hunting part of the seal harvest while women completed the butchering and other side duties—and situational—I tended to be "away" at school during the spring seal hunting season. Going seal hunting with my brother for this documentary was a new experience for me, but one that I naively thought I understood. In other words, growing up hearing seal hunting stories is not the same as being a mile out on the ice and completely dependent on understanding, predicting, and responding to spring ice conditions while barreling over snow-covered ice at 30+ miles per hour pulling a sled with 800 pounds of seals loaded in it.

During our hunting expedition, we rode out far onto the ice of the Kotzebue Sound and on our way out the ice seemed like it was good. We rode around until Qaluraq spotted groups of seals sunbathing in the distance. When he spotted them, he'd stop the snogo (snowmobile) a little way away from the seals and creep up on them by foot while I waited on our rig. Over the day, he successfully shot four adult seals. They were big animals and weighed about 200 pounds each and would be shared with multiple families in our community. As we rode on the ice with a sled heavy with seals, I filmed the white, snow-covered ice on the trail behind us breaking open and exposing a long, jagged line of black, freezing water. I trusted Qaluraq—I had to—but I also recognized that I was in extreme danger.

I yelled in his ear. I had to be loud to be heard through his fur hat and over the sound of the snogo's engine. "The ice is breaking behind us! It's too thin. I don't want to become a statistic. What are we going to do?"

Qaluraq simply responded, "We'll go fast," and kept driving us and our load of seals away from shore searching for a thicker area of sheet ice to safely drive on.

Because of this experience, when I hear stories that mention "the ice was thin," or "my trail kept breaking," or "we had to go fast," I understand the dangerous ice conditions those phrases imply. When I hear hunting stories that mention "the ice was good," it describes a different set of ice conditions. In scientific terms, "my trail kept breaking" means that a load that was less than one ton, consisting of an ~500-pound snowmobile carrying two people weighing about ~350 pounds together and pulling a sled carrying ~800 pounds of seals, was enough to break a section of the brackish ice in the Kotzebue Sound. According to the Alaska Clean Seas' *Ice Safety Awareness* booklet (n.d.), sea ice can bear less load than river ice (p. 24), and so the brackish nature of the Kotzebue Sound—a mixture of both sea water from the Chukchi Sea and river water from the Noatak and Kobuk rivers—makes it similar to, though less stable than, river ice (Fransson, 2009). River ice must be four inches thick to withstand a one-ton load (US Army Corps of Engineers, n.d., b). If river ice can bear a person on foot, it is at least two inches thick; if it can bear a snogo and a hunter, it is at least three inches thick (US Army Corps of Engineers, n.d., a). But if it fails once the hunter riding on a snogo is pulling a sled filled with seals weighing less than 2,000 pounds total, then the ice is less than four inches thick (see table 1.4).

In these tables I've presented thus far, I've extracted specific data from narratives using both context clues, such as I'yiiqpak equating the temperature −30°F to the air taking her breath away, and correlating published data with points in a narrative, such as ice load bearing information from the US Army Corps of Engineers and the approximate load described in the seal hunting story. Narratives are a way of communicating specific embodied knowledge and expertise about phenomena, such as weather or climate. Technical communication (TC) has a growing body of scholarship that places narrative as a theory-building methodology to "represent embodied knowledge" (Jones, 2017, p. 327), especially with regard to the lived experiences of members of multiply marginalized communities (e.g., Jones, 2016a, 2016b, 2016c, 2017, 2020). Similarly, Indigenous rhetorical practice and Indigenous inquiry also position narrative as theory and as a powerful teaching tool (e.g., Denzin & Lincoln, 2014; King et al., 2015; Riley Mukavetz, 2014; Powell, 2012). Narratives are a method of communicating important local expertise that is often overlooked as *scientific* knowledge in attempts to understand complex environmental situations. In the Arctic, the ice becomes a temporary extension of our land and is used heavily by members of our community. Studies related to the ways climate change is affecting the ice and the surrounding community's use of the ice can benefit by including user experience (UX) narratives. Bacha (2018) argues for the use of reflective storytelling in UX research, especially with a focus on "stories gathered during casual

Table 1.4. Seal Hunting Story

Narrative Reference	Ice Thickness (x) in inches	Notes
During our hunting expedition, we rode out far onto the ice in the Kotzebue Sound and on our way out the ice seemed like it was good.	$x \geq 3$	Ice held under ~850 lb. load: • snogo (500 lbs.) • two adults (350 lbs.)
When he spotted them, he'd stop the snogo a little way away from the seals and creep up on them by foot while I waited on our rig	$x \geq 2$	Ice held under ~200 lb. load: • man on foot (200 lbs.)
	$x \geq 3$	Ice held under ~650 lb. load: • snogo (500 lbs.) • one adult (150 lbs.)
As we rode on the ice with a sled heavy with seals, I filmed the white, snow-covered ice on the trail behind us breaking open and exposing a long, jagged line of black, freezing water.	$3 \leq x < 4$	Ice *broke* under ~1,650 lb. load: • snogo (500 lbs.) • two adults (350 lbs.) • four seals (800 lbs.)
"The ice is breaking behind us. It's too thin. I don't want to become a statistic. What are we going to do?"	$3 \leq x < 4$	Ice *broke* under ~1,650 lb. load: • snogo (500 lbs.) • two adults (350 lbs.) • four seals (800 lbs.)

conversations . . . those opinions are just as important as more formally collected sets of usability data" (pp. 190–200). In other words, including local UX of ice—especially informal utterances like "we had to go fast" or "the ice is real bad"—can complicate or complement how "science" understands ice.

My father, Caleb Lumen Pungowiyi, was a respected Siberian Yupik leader, climate change researcher, and advocate for considering Indigenous context-based knowledges as expert knowledge in scientific research. In a 2001 lecture at the University of Alaska–Fairbanks, he critiques western frames of "scientific" methods that exclude lived experience with climate change as data because it isn't scientifically measurable:

> One thing that . . . caught my eye . . . was "scientifically measurable programs." And I said oh my goodness . . . I think that we should not limit ourselves to scientifically measurable programs. To me, I think sometimes science is limiting . . . you don't get to see the whole world. And I think we should open the windows that give you the opportunity to see the whole world, and what that is, because sometimes science becomes a crutch that keeps you from proceeding or going forward.
>
> We know that the ice is . . . a lot thinner, it forms later, forming a lot later than what it used to in the 1950s, and it melts a lot easier. Maybe I shouldn't say "melts"—dissipates a lot easier—because "melts" means that it's melting with the heat of the sun and those kinds of things. But there are other factors that play into the dissipation of the ice, and as Native people or people who live on the coast, ice to us is kinda like supportive of life because it supports the seals, the fish, the birds, and the other things that are ice dependent, and it also becomes an extension of our land when it freezes over. We travel over it; we go fishing on it; it becomes, you know, part of our everyday life. But as it changes, though, it affects that opportunity for us to get out there. And a lot of you, if you were to talk to the hunters in springtime and the ice conditions are not ideal to them, "The ice is real bum this year," they'll say. But there's a deeper meaning to it in terms of what "bum" means, ok?
>
> We see ice in different "qualities." . . . If the winter is warm and ice doesn't freeze real solidly, it breaks up real easily in the springtime. Ice, when it forms, forms several ways. . . . Air temperature and water temperature play a very important role in the "quality" of ice that I talk about. As it gets colder the ice crystals squeeze, and they squeeze out this brine that's in the ice, the salt water. If it's not cold enough, some of that brine stays on the ice and it makes it really soft, easy to break up. And that's what I mean by a "quality" of the ice that we see. (Pungowiyi, 2001)

In this lecture, Pungowiyi demonstrates that there are many qualities about the ice that are contained within local Indigenous user observations, such as the utterance "the ice is real bum this year." TC as a field has historically leaned toward "scientifically measurable" data in its research, but as Pungowiyi argues, this focus is too narrow when it comes to accessing a nuanced understanding of issues like climate change. If Arctic researchers focus only on scientifically measurable data and continue to ignore the potential contextual data contained in local user narratives, then they are missing the opportunity to see "the whole world." Using the narrative extraction method I've developed and demonstrated in this chapter expands narrative's use as a form of technical communication that contains a wealth of useful data for researchers and can help guide climate action.

This narrative extraction method is useful in examining existing narratives, such as those provided earlier, as well as in the process of gathering new narratives. For example, if one was interviewing a hunter about a seal hunting experience on the ice and the hunter mentioned, "We had to go fast out there," this is a key moment where the interviewer should note that they need to ask a follow-up question regarding why they had to go fast. If the hunter then answers, "Because the ice was thin and kept breaking under our rigs," questions about the load (e.g., how many passengers, approximate weight of passengers, weight of load in sled) and about the date and location where this occurred would provide other useful data points. Obviously, asking the hunter if they knew the thickness of the ice and how they made that observation are other important follow-up questions. In my experiences hearing people telling stories about being out on the ice, multiple parallel observations are often described, such as mentioning the color of the ice, or mentioning water-filled snow on the ice, called *pukak*, or overflow, called *qaaptit*, or even physical descriptions of the ice's thickness from direct observation. All these narrative details, if noted and if the appropriate follow-up questions are asked, can yield useful quantitative data. Furthermore, once those data are correlated to a specific expression, they can be used as a reference point that the interviewer and the narrator can use to clarify subsequent details.

At this point it is important to state clearly that the data extracted from narratives in this manner do not necessarily produce exact measurements. Instead, they are a method of extracting data from stories at a relatively coarse granularity. These data can then be refined to a finer granularity by using follow-up questions or overlaying other correlating data (such as the ice load bearing data I used in my example). The more layers of correlating data, such as knowing the exact date of a hunting trip and then using that date to look up related weather conditions, the finer-grained data can become. The technical communication contained in narratives is a launching point to meaningfully incorporate these observations in research, much as Jones (2016a, 2016b, 2016c, 2017, 2020) has described. The use of narratives in scientific research is useful for enacting equity in methodology. For example, Jones (2016c) states that "narratives not only allow other voices and points of view to be heard and understood, but it pushes the researcher and scholar to examine his or her own positionality and enactment of power and agency in a reflexive manner. This acknowledgement of positionality can aid the technical communicator in embracing his or her political stance and enhancing agency in others in an attempt to bring about social change" (p. 351). Not only are narrative methodologies useful in incorporating the lived experiences of local experts, such as Inuit discussions of climate change's effect on their everyday lives, their use contributes to equity in research practices. Equity in this sense, contributes to a sense of cultural empowerment and allows participants to "guide the meaning-making process by sharing their own stories in their own voice" (Jones, 2017, p. 327). Further, narrative methodologies offer a useful way for Inuit communities to meaningfully contribute to research design, data collection, and data analysis and become full partners in research in their own communities.

Discussions of how worldview affects stories and their interpretation are important to consider, especially with regard to Inuit communities. For example, I've realized that western scientific/academic practice leans heavily on an anthropocentric worldview where humans are the ultimate actor in relationships with the world *around them*. In other words, humans are positioned as *the* point of view (POV). In my Inuit worldview, I rely heavily on a *kin*centric perspective—where human agency is tempered by, and in concert with, the agency of the world *in which* humans participate. In other words, humans are positioned as *a* POV. I do not consider human agency as the central, determining, or strongest vector in the web of interconnectedness of what is known as the world. It isn't the "world around us" at all; rather, it is, like my father states, the "whole world."

~

Reader, it is at this point that I again take up my letter and my story having done the work of demonstrating different ways of listening to my story and the stories of others to recognize the potential contained in what we are actually communicating. Here is where your work begins.

I have been researched. My family has been researched. We have given access to ourselves, our experiences, our ancestral knowledges, and our homes to fulfill others' ambitions, all in the hope that these actions can help our community. We have trusted others to depict our lives, experiences, knowledges, and communities fairly and accurately. We have lent our social capital to outsiders and vouched for them so that they may gain access into the homes of others in our community. We have incurred risk.

Tell me: What risk have you incurred?

~

In the summer of 1996, my family was hired to be field research assistants for a US Fish and Wildlife Service (USFWS) seabird survey on St. Lawrence Island in the Bering Sea off Alaska, where my dad Lumen is from. My sister Uula and I were hired as field assistants that summer and were trained to count crested auklets by the 100s with little handheld clickers. My mom I'yiiqpak was hired to be the camp cook and to keep everyone fed and comfortable. My father was a research partner in the project and also acted as a guide and a translator. He was key to the USFWS's access to people and lands on the island.

However, while we were there doing this research the joke on the island was the research itself. My father was teased by his family and by his friends because the people from the island *already knew* the kinds and numbers of birds that lived on the island. For example, my father told the USFWS that people on the island had already determined that the seabird population was roughly five million birds. However, the USFWS sent two western-trained scientists to supervise the bird counts on St. Lawrence Island to verify these observations. I remember sitting on a rock after being quickly trained to gauge what a 100-bird chunk looked like and then pressing the clicker for each 100-bird chunk I counted in the swirling dark cloud of thousands and thousands of birds above me. I clicked and clicked and hoped my chunking skills, gained after five minutes of training, were accurate. Islanders had been observing these swarms of birds for generations yet I, a college student with five minutes of training, was providing the "real" data. The irony of this situation was that my father spent his career trying to convince the scientific community to recognize the expertise of Indigenous peoples regarding their own lands and cultures. Regardless of my father's advocacy, there we were counting birds because the USFWS trusted my clicking abilities over the observations of the Siberian Yupik people living for millennia on St. Lawrence Island. It turns out there were indeed roughly five million birds (Stephensen et al., 1998), which was a win for my father's work to position local Indigenous observations as reliable, valuable, and accurate.

While my father never spoke in a negative way about this research experience on St. Lawrence Island, it privately became an inside joke between us. We used it to make lighthearted comments about what I came to understand as the tendrils of colonial scientific practice. For example, we might point out a gaggle of researchers arriving at the Kotzebue airport (zippy pants, lots of sturdy totes, and no hunting or kayaking gear). One of us might say to the other, "Maybe they're checking to see if we got five million birds," and we'd laugh. However, embedded in that simple exchange is a frustration that Indigenous knowledges are not recognized as equal to western scientific knowledge—even if Indigenous participation is an essential component to the success of western scientific research in Indigenous lands, waters, and communities. Jeannette Armstrong, Syilx/Okanagan Nation member and associate professor of Indigenous studies at the University of British Columbia, recently critiqued positioning western scientific frames as superior to Indigenous ones: "A general definition of Indigenous knowledge consists of those beliefs, assumptions, and understandings of non-western people developed through long-term associations with a specific place. . . . Therefore, Indigenous knowledge is considered the second tier of knowledge, that is, below science. This is racist" (Bonneau, 2021, para. 9–10). Indigenous knowledges are not secondary to western, "scientific" ones. Indeed, Inuit knowledges of our lands and waters are scientific, sovereign, and often sacred (Ellam Yua et al., 2022). These knowledges come from our communities observing, interacting, surviving together, and depending on the same lands and waters where our ancestors existed. Indigenous knowledges are at once intuitive, intergenerational, individual, communal, contextual, and complex. As I recently stated in a news article discussing the need for Indigenous representation in climate research in the Arctic, "[Western scientists] confirm what we already know, instead of investigating what we want to know. . . . It's taken tons of time for academia and the sciences to catch up to us. . . . They're just catching up" (Itchuaqiyaq quoted in Early, 2021, para. 12–14). Shifting one's worldview changes how things are done, valued, and enacted (Itchuaqiyaq, 2021). It is traditional for my people to have a kincentric worldview in their approach to understanding the "whole world," which is in direct contrast to approaches taken by western scientific method. This whole-istic knowingness, like understanding the way the weather feels and respecting the power of those conditions and how they affect you and everything else, is tapping into a deeper intuition we all have, but many ignore. My people are still very connected to that intuition because our lives *still* depend on it.

The Iñupiatun word for "environment" is *siḷa*. However, *siḷa* also has a broader meaning that is similar to a collective consciousness, whole-istic knowingness, or wisdom shared by everything. Canadian Inuk author Rachel Qitsualik describes it thus, "*Sila*, for Inuit, became a raw life force that lay over the entire Land; that could be felt as air, seen as the sky, and lived as breath" (Qitsualik quoted in Todd, 2016, p. 5). Data extracted from Inuit narratives demonstrate the power of *siḷa* for understanding our human impact, connections, and place in "the environment"—and for recognizing the multitudes of other "kin" and their impact, connections, and place in "the environment." *We* kin *are* the environment; we are both/and, the whole and a part. The recognition of this "whole world" intuitive knowledge is itself action that drives *reaction*. This connection I've described, in part, is communicated when Indigenous peoples say things like "we are part of the land," or "we come from the land." Our peoples in the Arctic are in a deep communication/interaction/being with *siḷa* and that influences our actions and reactions. Therefore, having Inuit in leadership roles is vitally important for research or climate change policy activities, and so on, that aim to understand and generate action plans for the Arctic. The intuition related to *siḷa*, and the ability to recognize and understand that intuition, is included in terms like "Indigenous knowledges."

Unfortunately, while Indigenous knowledge is considered valuable, *how* it is incorporated in research or policy reflects institutional bias (Ellam Yua et al., 2022). Indigenous knowledges are not typically considered "scientific" but are instead considered "cultural," or "primitive," basic observations. However, in reality, there is often rich, data-oriented detail contained in mundane utterances or observations about lived experiences in the Arctic. Because of this institutional bias, Inuit intuition/bodily knowledges are not typically used in developing policy, action plans, or research plans related to climate change monitoring or response.

Indigenous rhet/comp scholar Regina McManigell Grijalva (2020) argues that ethical storytelling and storylistening practices must pay particular attention to contexts and how they affect the communication process. McManigell Grijalva explains: "An important concept to understanding indigenous communication is that the sacred, the spiritual, and the physical are often conceived of as unified and not easily separated. . . . The coherence of a context with all its related reasons, values, and actions will illuminate the interconnectedness of their underlying ethics" (p. 33). In other words, while I'm trying to teach you a method of extracting data from Inuit narratives, I'm also trying to convey a bigger message: Inuit living in the Arctic have unique expertise about climate change and its a/ effect on communities, landscapes, waters, ice, plants, and animals. This same message was conveyed by my father and others decades ago and has been continually expressed by Inuit leaders and scholar-allies since that time (e.g., Krupnik & Jolly, 2010, Ellam Yua et al., 2022). Henry Huntington, a non-Indigenous scholar-ally and close friend of my father describes the importance of Indigenous contributions to scientific knowledge: "Researchers need to understand the implicit method that lies beneath the knowledge of Arctic peoples, just as they work diligently to follow the scientific method in other research. Only in this way can researchers appreciate not only the information that is generated by the knowledge of the Arctic residents, but also their perspective on the environment, their relationship with it, and what if any actions are needed to protect that relationship" (Huntington, 2010, p. xxxii). Inuit scientific observations are not second tier to observations made via the scientific method. Rather, Inuit expertise is essential to research within Inuit communities and their related lands and waters. Thus, Inuit should lead these research efforts rather than fulfill subordinate, supportive roles, such as merely acting as research participants relegated to supplying access or narratives. Yes, our stories are important, and I argue that they provide important data. But *how* can that data be properly understood, analyzed, and integrated into findings and conclusions without active Inuit leadership, expertise, and analysis in such research? It is interesting to note that these arguments for a reconsideration of how scholars approach knowledge and research with/in Inuit communities echo existing critiques of service-learning partnerships or community-engaged learning projects between universities and organizations (e.g., Batova, 2020; Kimme Hea & Wendler Shah, 2016; Scott, 2004).

∾

Reader, there is a lot to unpack in this story I've been attempting to tell you. I've told you about my family and our activities and how those activities have helped us understand the "whole world" around us. I've told you about our adventures and our frustrations. I've provided examples of using narratives as potential scientific data sources. I've described a method of narrative data extraction and indicated how overlaying complementary data sources, such as ice load data, and listening for potentially rich data contexts, such as "we had to go fast," can help to refine such data. Ultimately, I've argued that methods like narrative data extraction in Inuit communities require a contextual understanding of the situation from which these narratives arise in order to successfully apply these methods. Thus, narrative methodologies require equitable local partnerships at a leadership level in the data collection, in the data analysis processes, and in the research design itself. Inuit partnerships enrich research regarding climate change in Arctic landscapes as well as inform climate action to protect those landscapes. We Inuit have been taught to value and develop our relationship with the land, the waters, and the ice that surround us. We have been taught to listen to its stories.

Reader, I'll end my narrative with a story my dad told our family many times and that shaped how I think about the world in profound ways:

> I was [raised by my grandmother]. . . . And one of the things I remember is one time we went camping by ourselves in the summertime. I was about seven, eight years old. . . . We went out [to camp because] she wanted to gather greens for the winter. So, we went over there just by ourselves, just me and my grandmother, and she'd do the picking and I'd do the wandering around exploring, but my job was to pack that bag of greens back to the camp. And she would take a break once in a while, and one time I remember her telling me, "Son, listen to the whispers of the grass," in our language, "and it'll tell you stories of ages long past." And I used to wonder what that meant. What did she mean by "listen to the whispers of the grass"? I've thought about it over the years, and I think that if we don't listen, we don't hear. . . . And to me the whispers of the grass has meant a lot to me, to listen to the whispers of the grass. (Pungowiyi, 2001)

Quyanaqpak for listening.

References

Alaska Clean Seas. (n.d.). *Ice safety awareness: A practical guideline to ice safety.* Retrieved January 28, 2021, from https://archive.epa.gov/emergencies/content/fss/web/pdf/karellaice.pdf

Bacha, J. A. (2018). Mapping use, storytelling, and experience design: User-network tracking as a component of usability and sustainability. *Journal of Business and Technical Communication, 32*(2), 198–228. https://doi.org/10.1177/1050651917746708

Batova, T. (2020). An approach for incorporating community-engaged learning in intensive online classes: Sustainability and lean user experience. *Technical Communication Quarterly.* https://doi.org/10.1080/10572252.2020.1860257

Bonneau, A. (2021, January 31). *Racism at root of Indigenous knowledge being ignored says UBC professor.* Retrieved January 31, 2021, from https://www.aptnnews.ca/national-news/racism-at-root-of-indigenous-knowledge-being-ignored-in-academia-says-ubc-professor/

Denzin, N. K., & Lincoln, Y. S. (2014). Introduction: Critical methodologies and Indigenous inquiry. In N. K. Denzin, Y. S. Lincoln, & L. T. Smith (Eds.), *Handbook of critical and Indigenous methodologies* (pp. 1–20). Sage. https://doi.org/10.4135/9781483385686.n1

Early, W. (2021). *More than 200 researchers sign letter requesting more Indigenous input in national Arctic science initiative.* Retrieved February 8, 2021, from http://kotz.org/2021/02/05/more-than-200-researchers-sign-letter-requesting-more-indigenous-input-in-national-arctic-science-initiative/

Ellam Yua, Raymond-Yakoubian, J., Daniel, R., & Behe, C. (2022). A framework for co-production of knowledge in the context of Arctic research. *Ecology and Society, 27*(1), 34. https://doi.org/10.5751/ES-12960-270134

Fransson, L. (2009). *Ice handbook for engineers: Version 1.2.* Luleå Tekniska Universitet Institutionen för Samhällsbyggnad. Retrieved January 28, 2021, from http://static1.1.sqspcdn.com/static/f/572109/15210238/1321785520853/ice_handbook_2009-1.pdf?token=MLeOBImQQaCC44UZ8ih3NG5rZC8%3D

Huntington, H. P. (2010). Preface: Human understanding and understanding humans in the Arctic system. In I. Krupnik & D. Jolly (Eds), *The earth is faster now: Indigenous observations of Arctic environmental change* (pp. xxix–xxxv). Arctic Research Consortium of the United States.

Itchuaqiyaq, C. U. (2021). Iñupiat Iḷitqusiat: An Indigenist ethics approach for working with marginalized knowledges in technical communication. In R. Walton & G. Agboka (Eds.), *Equipping technical communicators for social justice work: Theories, methodologies, and pedagogies.* Utah State University Press.

Jones, N. N. (2016a). Found things: Genre, narrative, and identification in a networked activist organization. *Technical Communication Quarterly, 25*(4), 298–318. https://doi.org/10.1080/10572252.2016.1228790

Jones, N. N. (2016b). Narrative inquiry in human-centered design: Examining silence and voice to promote social justice in design scenarios. *Journal of Technical Writing and Communication, 46*(4), 471–492. https://doi.org/10.1177/0047281616653489

Jones, N. N. (2016c). The technical communicator as advocate: Integrating a social justice approach in technical communication. *Journal of Technical Writing and Communication, 46*(3), 342–361. https://doi.org/10.1177/0047281616639472

Jones, N. N. (2017). Rhetorical narratives of Black entrepreneurs: The business of race, agency, and cultural empowerment. *Journal of Business and Technical Communication, 31*(3), 319–349. https://doi.org/10.1177/1050651917695540

Jones, N. N. (2020). Coalitional learning in the contact zones: Inclusion and narrative inquiry in technical communication and composition studies. *College English, 82*(5), 515–526.

Kimme Hea, A. C., & Wendler Shah, R. (2016). Silent partners: Developing a critical understanding of community partners in technical communication service-learning pedagogies. *Technical Communication Quarterly, 25*(1), 48–66. https://doi.org/10.1080/10572252.2016.1113727

King, L., Gubele, R., & Anderson, J. R. (Eds.). (2015). *Survivance, sovereignty, and story: Teaching American Indian rhetorics.* Utah State University Press.

Krupnik, I., & Jolly, D. (2010). *The earth is faster now: Indigenous observations of Arctic environmental change.* Arctic Research Consortium of the United States.

NPR. (2009, January 24). *Why does cold weather cause runny noses?* NPR. Retrieved January 5, 2022, from https://www.npr.org/templates/story/story.php?storyId=99844567

McManigell Grijalva, R. (2020). The ethics of storytelling: Indigenous identity and the death of Mangas Coloradas. *College Composition and Communication, 72*(1), 31–57.

Powell, M. (2012). 2012 CCCC Chair's address: Stories take place: A performance in one act. *College Composition and Communication, 64*(2), 383–406.

Pungowiyi, C. L. (2001). December 2, 2001, lecture about climate change at University at Alaska, Fairbanks. Retrieved from https://jukebox.uaf.edu/site7/interviews/3551

Riley Mukavetz, A. M. (2014). Towards a cultural rhetorics methodology: Making research matter with multi-generational women from the Little Traverse Bay Band. *Rhetoric, Professional Communication and Globalization, 5*(1). Retrieved from http://www.rpcg.org/index.php?journal=rpcg&page=article&op=view&path%5B%5D=98%5Cnpapers3://publication/uuid/22606F13-24D8-4F4D-B2BA-182A1E5FD5E3

Scott, J. B. (2004). Rearticulating civic engagement through cultural studies and service-learning. *Technical Communication Quarterly, 13*(3), 289–306. https://doi.org/10.1207/s15427625tcq1303_4

Simonee, N., Alooloo, J., Carter, N. A., Ljubicic, G., & Dawson, J. (2021). Sila qanuippa? (How's the weather?): Integrating Inuit qaujimajatuqangit and environmental forecasting products to support travel safety around Pond Inlet, Nunavut, in a changing climate. *Weather, Climate, and Society, 13*(4), 933–962.

Stephensen, S. W., Pungowiyi, C., Mendenhall, V. M. (1998). A seabird survey of St. Lawrence Island, Alaska, 1996–1997. US Fish and Wildlife Services Report, Migratory Bird Management (MBM), Anchorage, AK.

Todd, Z. (2016). An Indigenous feminist's take on the ontological turn: "Ontology" is just another word for colonialism. *Journal of Historical Sociology, 29*(1), 4–22. https://doi.org/10.1111/johs.12124

US Army Corps of Engineers. (n.d., a). *Ice thickness and strength for various loading conditions.* Retrieved January 28, 2021, from https://rivergages.mvr.usace.army.mil/WaterControl/Districts/MVP/reports/ice/ice_load.html

US Army Corps of Engineers. (n.d., b). *Safety on floating ice sheets.* Retrieved January 28, 2021, from https://rivergages.mvr.usace.army.mil/WaterControl/Districts/MVP/reports/ice/safety.html

Wells, J. K. (1974). *Ipani Eskimos: A cycle of life in nature.* Alaska Methodist University Press.

Chapter 2

Boundary Waters

Deliberative Experience Design for Environmental Decision Making

DANIEL CARD

As this collection demonstrates, many technical communication (TC) researchers and practitioners increasingly feel drawn to environmental action. Indeed, whether lead contamination in drinking water, oil spills, or the climate crisis, we find ourselves compelled to *do something*—to analyze and critique but also find ways to directly intervene. I see this push for action, engagement, or intervention as emergent from concerns about the places we live, work, and play, and influenced heavily by TC's long-standing commitment to user advocacy and more recent embrace of social and environmental justice frameworks (Cagle, 2017; Sackey, 2018; Walton et al., 2019).

At the same time, technical communicators and allied scholars also recognize that environmental problems are wicked problems—problems that involve multiple interacting systems, uncertainty, and value-laden solutions (Blythe et al., 2008; DeVasto et al., 2019). Given the sociotechnical nature of wicked problems, equitable solutions do not arise from technical expertise alone. Each possible solution will affect various stakeholders differently, and that makes these problems "public" in an important sense. In environmental policy contexts, recognizing a problem as wicked requires us to figure out how to bring stakeholders to the table—not only those with expertise in geology, ecology, economics, and so on, but also those impacted by the problems, solutions, and alternatives under consideration. Public participation in environmental policymaking represents one key site in which this work happens. Technical communicators bring a compelling set of skills and commitments to this work, but as I argue in this chapter, fostering equitable decision making (i.e., procedural justice) is itself a wicked problem (Card, 2020; Walker, 2010). Common approaches, though well intentioned, tend to produce deliberative winners and losers as well as policy winners and losers, often reproducing preexisting inequalities in the process.

This wicked nature of environmental problems is certainly on display in the controversial plan to build a mine just outside the most visited wilderness area in the nation: the Boundary Waters Canoe Area (BWCA). While most visitors think of the BWCA as home to pristine water and abundant wildlife, it also sits atop one of the world's largest untapped mineral deposits—over four billion tons of copper, nickel, and other precious metals. In December of 2019, the mining company Twin Metals Minnesota submitted a formal plan to develop an underground mine within the same watershed as the BWCA, triggering a multiyear state and federal environmental review process. Technical communicators invested in environmental action must attend to the policy apparatus—to the dance of legislation, regulation, and litigation that often marks the success or failure of environmental efforts. The environmental review of the Twin Metals Project (hereafter, the TMP), which is unfolding at the time of writing, provides a useful opportunity to explore the policy process as a site of environmental action. While it is common in our journals and collections like this one to read reports on completed work or retrospective analyses of how a case went wrong, prospective work—the work that must precede effective intervention—is often occluded.

In this chapter, I try to make visible the labor required to navigate environmental action in policy spaces. Rather than draw on a completed case to make an argument about what should have been done, I examine agency websites, news articles, issue-based task force reports, environmental assessments, industry and advocacy communication, and governmental policies and guidance to consider the ways technical communication research and practice might intervene going forward.

This chapter proceeds in two parts. In the first half, I suggest that TC researchers have productively positioned themselves as citizen advocates in policy spaces traditionally dominated by experts. I then draw on the conflict surrounding Twin Metals to situate the environmental review and associated public participation processes as critical moments of environmental action in which a range of publics must navigate the technical and political. In doing so, I suggest that TC explore alternative frameworks that explicitly engage the normative dimensions of policy.

In the second half, I suggest that environmental justice represents one such framework, but that few in TC have explored the ways agencies conceptualize it. I begin that work by examining government commitments to environmental justice at the state and federal level. In an effort to consider how an environmental justice framework might inform the impending environmental review of the TMP, I draw on the recent environmental review of a similar mine.

TC as Advocate for Access

Recognizing that policy decisions often directly impact the places we live, work, and play, scholars in TC and allied fields have tried to make good on their commitment to user advocacy by promoting citizen agency in policy contexts. In some cases, this has meant promoting access to information in policy spaces (Jones et al., 2012; Jones & Williams, 2017; St. Amant, 2018). As St. Amant suggests, access to information involves two interlocking parts: "The first is availability—that is, the sources of information are available for us to consult as needed. . . . The second concept is comprehensibility—or information is presented in a way we can easily understand and act on" (p. xxi). This positions technical communicators, who are skilled in working with highly specialized discourse and often versed in best practices for plain language and accessibility—as ideal individuals to foster advocacy by designing materials and conveying highly technical information to a range of audiences.

Recognizing the difficulty of acting on information in policy spaces, TC scholars have also sought to understand, change, and in some cases directly participate in public participation processes (Moore & Elliott, 2016; Moore, 2017a; Moore, 2017b; Simmons, 2008). For example, Blythe et al. (2008) illustrate how technical communicators can identify and support the interventional strategies of citizens and Moore (2017b) articulates a vision of the technical communicator as participant, facilitator, and designer in public engagement projects. Whether improving access to information or to the processes and spaces in which policy decisions are made, technical communication researchers have often positioned themselves as advocates for the community in spaces that are traditionally dominated by experts or unduly influenced by industry. This approach has been productive, especially in cases where the impacts of a project are limited to a small, specific locale and the community/public/citizens are united in their concerns. However, many of the most pressing environmental challenges and their proposed solutions have impacts that ripple across geographies, are marked by uncertainty, and are often polarizing along political lines (Cagle & Herndl, 2019). As I illustrate in the next section, the wicked nature of environmental problems is compounded by the complexity of the policy process. Taken together, this presents significant challenges to technical communicators as they try to figure out for whom or what they should advocate and how they might do so effectively.

Situating Environmental Review of the TMP

On December 18, 2019, the Minnesota Department of Natural Resources announced that it had received a formal project plan from Twin Metals Minnesota (hereafter, Twin Metals). In the plan, Twin Metals provides details on its proposed underground mine as well as facilities to process minerals and store waste. Per both the Minnesota Environmental Policy Act (MEPA) and National Environmental Policy Act (NEPA), the proposal requires state and federal agencies to prepare an environmental impact statement (EIS) to evaluate the potential environmental consequences of a proposed project. While it is common in cases like this for relevant federal agencies to collaborate with state agencies on a single, shared EIS, the Minnesota Department of Natural Resources (hereafter DNR) announced that it will prepare an independent, state-only EIS for the project, citing the importance of a transparent, neutral, and science-based review. The Department of the Interior and Bureau of Land Management (BLM), then, will conduct its own EIS for the project (Bureau of Land Management [BLM], 2020).

In this announcement, the DNR echoes language used by both critics and proponents of the mine. Twin Metals, environmental advocates, and both Democratic and Republican representatives have all argued publicly that the regulatory process should be "based on science, not politics." The DNR maintains this focus on neutrality, articulating that the purpose of an EIS is to "provide project decision-makers and the public with objective facts about the potential for significant environmental, social, and economic effects of a proposed project" (Minnesota Department of Natural Resources [DNR], 2020). Per both NEPA and MEPA, an EIS must include the action's description; environmental, social, and historic impacts; considered alternatives; potential mitigation measures; and irreversible commitments of resources (Minnesota Environmental Quality Board & Seuffert, 2015).

The completion of the environmental impact statement paves the way for Twin Metals to apply for the permits required to construct and operate a mine. It is worth noting that EISs are advisory in nature; NEPA requires agencies to take a hard look at the environmental impacts of government actions but does not prevent them from taking those actions. The identification of significant impacts does not preclude the approval of project permits. As argued in an oft-quoted Supreme Court decision, "NEPA merely prohibits uninformed—rather than unwise—agency action" (*Robertson v. Methow Valley Citizens Council*, 1989). The company already obtained a lease for the land in the Superior National Forest where the mine would be constructed, the renewal of which was granted in 2019 by the Trump administration's BLM in a decision that reversed the 2016 denial by the Obama administration's BLM. State and local agencies can only issue Twin Metals the permits required to construct the mine after the DNR determines that the EIS is *adequate*—that it accurately and completely evaluates the potential effects of the proposed project. Those agencies would then rely on the EIS as they decide not only whether to grant the permit but also what provisions (e.g., pollution requirements) and conditions (e.g., monitoring and reporting requirements) to include.

Table 2.1, while not exhaustive, illustrates some of the agencies and documents that the TMP must navigate. In each case, the public will have the opportunity to participate in some capacity. The first formal opportunity for public participation in the review of the mine proposal will happen at some point during the current phase of the EIS process: scoping. According to the Minnesota DNR, the purpose of scoping is "to identify the potentially significant environmental and socioeconomic issues requiring detailed analysis, the alternatives to be evaluated, and the potential mitigation options for the EIS" (Minnesota DNR, 2020). In other words, during the scoping phase, 1) issues are deemed relevant and significant and 2) alternatives to be evaluated are identified. As such, the scoping phase plays a critical role in determining what questions are relevant and what data will be gathered, essentially framing the entire review process. At the time of writing, the Minnesota DNR has received the Twin Metals proposal and is preparing the scoping environmental worksheet, which serves as a detailed outline for the first draft EIS (see table 2.2).

Table 2.1. Regulatory Pathway for Twin Metals Project

Proposed Action	Explore land with the intent to develop a mine	Apply for the air, water, etc., permits required to operate a mine	Develop and operate a mine
Regulating Agency	Bureau of Land Management, US Forest Service	"Regulating Governmental Unit"—Minnesota Department of Natural Resources, Bureau of Land Management	Minnesota Department of Department of Natural Resources, Minnesota Pollution Control Agency
Critical Documents	Mineral lease, prospecting permit	Mine proposal, environmental impact statement	Operation permits

Table 2.2. Timeline: Minnesota DNR's Environmental Review of the Twin Metals Project

Phase	Timeframe
Pre-project	
Proposal submitted	Completed
Scoping	
Prepare scoping environmental assessment worksheet (EAW) and draft scoping decision	Variable
Public comment period and public meeting	Minimum of 30 days
Final scoping decision	Minimum of six weeks
EIS Preparation	
Prepare draft EIS	Variable
Public comment period on draft EIS and public meeting	Minimum of 45 days
Final EIS preparation	Variable
Final EIS public comment period	Minimum of two weeks
Determination of EIS adequacy	Variable

The Minnesota DNR has also issued a request for proposals to hire a consultant to assist in their review of the TMP, a frequent practice when the review at hand is exceptionally complex, controversial, or otherwise likely to get significant public attention.

The EIS as an Occasion for Political Dialogue

This environmental impact statement will be exceptionally complex and controversial. What are the potential environmental impacts of this mine? Even in its narrowest configuration, this itself is a complicated question to answer, requiring expertise in a range of specialized fields including but not limited to limnology, hydromorphology, geomorphology, mineralogy, engineering, ecology. Yet I suspect lawyers, ecologists, and technical communicators alike recognize that despite best attempts at neutrality, descriptions of these documents are always political. They are political in the sense that they will become leveraged in political discourse, but they are also political because they contain an implicit expression of values.

In the case of the TMP, the dominant discourses that circulate in the major state and local news about the proposed mine provide insight into the political nature of environmental review. The most reductive "two-sides" version of the controversy surrounding the BWCA and Twin Metals Project pits 1) environmentalists who are trying to protect the BWCA from the threat of a polluting industry against 2) blue-collar workers in the economically depressed northern Minnesota communities of Ely (pronounced EE-lee) and Babbitt. Within this debate, the "Campaign to Save the Boundary Waters" is frequently featured in state and local news coverage. Led by "Northeastern Minnesotans for Wilderness," the campaign is "organized by local residents in and around Ely . . . dedicated to creating a national movement to protect the clean water, clean air and forest landscape of the Boundary Waters Canoe Area Wilderness and its watershed from toxic pollution caused by mining copper, nickel and other metals from sulfide-bearing ore" (Northeastern Minnesotans for Wilderness, 2021).

The campaign, in concert with 25 partners at national (e.g., Izaak Walton League, American Canoe Association, and League of Conservation Voters) and local levels (e.g., Sportsmen for the Boundary Waters, Minnesota Conservation Federation) has amplified a series of talking points via interviews and other advocacy activities that focus primarily on the potential threat sulfide mining poses to the nearby BWCA.

These campaigns work to funnel people to the formal opportunities for public participation required by laws like NEPA. For example, comments submitted to the Bureau of Land Management (BLM) in 2019 when Twin Metals renewed its lease mirror the campaign's talking points. Most commenters opposed the lease renewal out of concern for potential environmental impacts to the Rainy River Watershed and more specifically BWCA. Readers familiar with Herndl and Brown's *Green Culture* (1996) or Ross's *Topic-Driven Environmental Rhetoric* (2017) will recognize many of the environmental commonplaces identified in those texts as present also in the anti-TMP discourse. Commenters spoke of the BWCA as a *pristine, untouched wilderness* that represents one of the only remaining places where we can *renew our spiritual connection to nature*. They emphasized the obligation we have to protect the BWCA for *future generations*, the *inherent risk* of sulfide mining, and the environmental track record of the *foreign conglomerate* Antofagasta, the Chilean company of which Twin Metals is a subsidiary. They expressed anger that the federal government is selling resources that the people of Minnesota want protected and that the federal government is prioritizing profit over science and the will of the people.

A smaller but still significant number of comments in support of the lease renewal mirror talking points from Twin Metals' public relations efforts. Specifically, those comments argue the mine would create an estimated 700+ jobs in an economically depressed community that barely gets by on a small recreation economy; the mine would feature state-of-the-art technology to ensure environmental protection and the minerals are needed for electronics and green technology, like solar panels and batteries for electric cars; "we" should mine metals here in Minnesota where there are strong environmental and labor protections; and finally, the liberals from the metro who oppose the mine should focus their attention on the environmental disaster that is the Twin Cities.

These talking points work in part to construct two distinct problems. In one formulation, an inherently toxic process poses an imminent threat to a sacred place, benefiting only a greedy, foreign corporation. In another, advances in technology allow us to simultaneously protect the environment, create much-needed jobs, and secure critical resources in the race against the climate crisis. In these contrasting formulations, this environmental conflict quickly grows to include poverty and climate change (both wicked problems themselves). Automation, industrialization, green technology, child labor, and science denial are invited into the fray.

As these competing efforts to construct the problem illustrate, environmental problems are characterized by not only technical but also social and political complexity. Even something as seemingly simple as describing the location of the mine shifts the scope of the problem in critical ways. For example, the mining interests often describe the project as located between the cities of Ely and Babbitt and add that the area has long been sustained by mining. This description conjures a somewhat different set of concerns and stakeholders than the description offered by environmental interests that locate the project within the Boundary Waters Canoe Area accompanied by descriptors that reference the area's importance as a wilderness and natural preserve.

In addition to illustrating some of the politics that those facilitating public participation and writing the EIS will need to navigate, these discourses challenge TC's impulse to advocate for the local community in the face of expert or industry overreach. Even if we bound "community" to the city limits of Babbitt and Ely (the two small towns nearest the proposed mine), we find a range of publics invested in the TMP, including those who would like to see mining jobs available for themselves or their children as well as the owners and employees of the recreation outfitters that make up much of the current local economy. Certainly, this is in part due to a concerted effort by advocates for and against the TMP to occupy the position of community. Indeed, community represents a significant topos in discourses of environmental conflict. As Raymond Williams says, "Community can be the warmly persuasive word to describe an existing set of relationships, or the warmly persuasive word to describe an alternative set of relationships. What is most important, perhaps, is that unlike all other terms of social organization (state, nation, society, etc.) it seems never to be used unfavorably, and never to be given any positive opposing or distinguishing term" (1985, p. 76). It's clear the TMP feels significant to people far beyond the project site, considering around 40,000 people submitted comments while combined Ely and Babbitt have a population of fewer than 5,000 people. And even if Ely and Babbitt were univocal in their support, should we then ignore the concerns of communities across the state who are invested in the area? The project site is within the Superior National Forest and in proximity to the BWCA. As such, any light, noise, water, or air pollution from mine operation may affect publics far beyond the borders of Ely and Babbitt.

It's unsurprising that environmental advocacy organizations and industry stakeholders invest significant energy trying to shape the environmental review process, given that the EIS provides agencies the basis on which to approve or deny a permit and add critical restrictions. These organizations understand that it will be much easier for the Minnesota Department of Natural Resources to deny a water permit if there is documentation that many federally protected species would be threatened if the permit were granted. It is tempting to disregard these dominant discourses as the result of sophisticated marketing efforts that overstate the potential for pollution or benefits to the green economy; however, in the long run public participation is at least in part about building trust and understanding between agencies and the people they serve, so ignoring narratives that resonate comes with its own risks.

As I look toward the environmental review of the TMP, this brief examination of the regulatory context and political discourse provides insight into the challenges of environmental action in policy contexts. While TC has generally positioned itself as citizen or community advocate, it's unclear in the TMP for whom or to what end we might advocate. In amplifying some voices, there is always a risk of inadvertently silencing others or setting up false equivalencies. Considering this tension, TC should explore alternative frameworks that explicitly engage the normative dimensions of policy processes and outcomes. In what follows, I explore environmental justice as one framework to guide TC efforts in policy spaces.

Toward Environmental Justice in the TMP Review

Environmental action is inextricably bound to questions of equity and justice, so scholars working in this area have much to learn from the growing body of work in technical communication that engages decoloniality (Agboka, 2014; Eichberger, 2019; Itchuaqiyaq & Matheson, 2021), social justice (Walton et al., 2019), and feminisms (Haas & Frost, 2017). Consonant with the broader turn toward issues of equity, some scholars in TC and allied fields have engaged environmental justice (EJ) to center and elevate the voices of those who are excluded from decision-making processes about their own lives (Enriquez-Loya & Léon, 2020; McGreavy et al., 2020; Sackey, 2018, 2020). The environmental justice movement emerged in the 1960s as people of color sought to address the inequity of environmental protection in their communities. In 1991 at the First National People of Color Environmental Leadership Summit, several hundred people who identify as Native American, African American, Latino, and Asian Pacific adopted 17 environmental justice principles that continue to guide the movement. As Sackey (2018) notes, these principles emphasize "meaningful involvement of people in the development, implementation, and regulation of environmental policy . . . requiring that research and inquiry be a collaborative endeavor that reaches across communities and draws upon various situated knowledges" (p. 150). Infusing the review process with environmental justice values—transparency, participation, accountability, sovereignty, relationality—requires attention to issues of power, privilege, and positionality. In short, environmental justice asks us to consider whether an environmental review process fosters dialogue among the right people, at the right time, to the right end.

This framework should be particularly compelling to TC in part because agencies are already supposed to be using it. The DNR's environmental review process and any potential permits that follow *should* be informed by an EJ framework per an executive order at the federal level and a series of agency-specific commitments and guidance documents that followed. More specifically, Executive Order 12898 directs all federal agencies to make environmental justice part of its mission by identifying and addressing "disproportionately high and adverse human health or environmental effects of its programs, policies, and activities on minority populations and low-income populations" (Exec. Order No. 12898, 1994). The Presidential Memorandum accompanying this 1994 directive from then-president Bill Clinton emphasizes the importance of using the NEPA process to promote environmental justice. In the wake of the memo, an Environmental Justice Interagency Working Group was formed and subsequently established a NEPA committee in 2012 to improve the consideration of environmental justice issues in the NEPA process (US Environmental Protection Agency, 2016). More recently, President Biden pledged to reinvigorate Clinton's 1994 directive through revisions that elevate environmental justice (*The Biden Plan*, 2020). At the state level, the Minnesota Pollution Control Agency has developed its own guidance that prioritizes environmental justice. Table 2.3 summarizes these environmental justice resources.

Per EO 12898 and subsequent federal guidance documents, "meaningful involvement" of minority and low-income populations is a critical part of fulfilling environmental justice commitments in NEPA reviews. Indeed, the 2016 report *Promising Practices for EJ Methodologies in NEPA Reviews* outlines a series of guiding principles and specific steps to promote meaningful involvement (pp. 6–7):

Table 2.3. Resources on Environmental Justice in the Context of NEPA Reviews

Resource	Description
Executive Order 12898	Directs federal agencies to make environmental justice part of its mission by "identifying and addressing, as appropriate, disproportionately high and adverse human health or environmental effects of its programs, policies, and activities on minority populations and low-income populations," including tribal populations
Promising Practices for EJ Methodologies in NEPA Reviews	Provides guidance and practices to enhance EJ considerations in the context of NEPA reviews
Community Guide to Environmental Justice and NEPA Methods	Provides information to help communities ensure EJ issues are adequately considered in NEPA reviews
Environmental Justice; Guidance under NEPA	Outlines six principles to guide consideration of EJ
MPCA Environmental Justice Framework	Commits MPCA to fair treatment and meaningful involvement, per EJ principles
MPCA interactive map of EJ areas of concern	Tool to identify whether a project may have adverse impacts on communities of color, low-income communities, or communities within tribal boundaries
EJSCREEN	EPA's environmental justice mapping tool

- Consider "adaptive and innovative approaches to both public outreach (i.e., disseminating information) and participation (i.e., receiving community input) since minority and low-income populations often face different and greater barriers to participation."

- Meaningful engagement efforts with potentially affected populations are generally most effective when "initiated early and conducted often."

- Meaningful engagement efforts "can play an important role in . . . [collecting] data used to inform the decision-making process."

- "Convening project-specific community advisory committees and other established groups to identify potential impacts and mitigation measures (as part of the NEPA review process) comprised in part of potentially affected minority populations and low-income populations can enhance agencies' understanding of the proposed action's potential impacts and alternatives, and can be a valuable public participation strategy."

- "Providing minority populations and low-income populations, and other interested individuals, communities, and organizations with an opportunity to discuss the purpose and need statement early in the NEPA process can help focus public input. Explaining the purpose and need for agency action to the minority populations and low-income populations early in the NEPA process can help focus meaningful engagement (i.e., public outreach and participation) efforts."

This guidance aligns with TC's focus on procedural justice insofar as it works to create a space in which those impacted can participate in decision making, but it also usefully extends it by prioritizing those who face the most barriers, will be most impacted, or have historically been marginalized in and by these processes. The emphasis on *meaningful* participation in these guiding principles also underscores the need for alternatives to the conventional "notice and comment" approach to public participation. In the next section, I look back to recent moments of participation to consider whether conventional approaches meet the demands of environmental justice.

Recent Examples of Superficial Inclusion

Although the DNR is compelled by federal guidance to pursue EJ in their review of the TMP, it is unclear how they will approach it given the lack of agency statements that clearly outline action steps. Mapping the environmental review process and exploring environmental justice resources clarifies what an equitable process might look like: who should be involved, when, and in what capacity. Unfortunately, recent history does not paint a promising picture. In 2005, PolyMet Mining, another national mining company much like Twin Metals, submitted to the DNR a plan to build the "NorthMet Mine," an open-pit sulfide ore mine in the Superior National Forest. The NorthMet Mine is similar in many respects to the Twin Metals mine that is just beginning the environmental review process: both are in the Superior National Forest in St. Louis County; both would be sulfide mines.

Documentation from the review of the NorthMet mine—specifically, appendix D of the 2009 Preliminary Draft Environmental Impact Statement (PDEIS)—provides important context for the impending review process of the TMP (Minnesota Department of Natural Resources, 2009, pp. 319–847). The PDEIS for the NorthMet mine was prepared by MN DNR's review contractor and subsequently reviewed by MN DNR, coordinating agencies, and Tribal Cooperating Agencies, which included the Bois Forte band of Chippewa, Fond du Lac band of Lake Superior Chippewa, Grand Portage Band of Chippewa, Great Lakes Indian Fish and Wildlife Commission, and the 1854 Treaty Authority. The Tribal Cooperating Agencies submitted position statements in the form of blue annotations on the PDEIS—about 528 pages of draft EIS with annotations inserted throughout. These insertions rejected claims about impacts based on the use of unreliable models or inadequate data; contested DNR interpretation and application of definitions (e.g., use of the term "waters of state"); and noted the lack of attention to impacts that are particularly relevant to them, for example, air and water quality impacts to territories where they maintain rights to harvest natural resources. Perhaps most relevant when considering the DNR's ongoing scoping for the TMP, numerous commentors took issue with the way they were involved—at what point during the process and in what capacity. For example, on the "Cultural Background" section, the Tribal Cooperating Agencies wrote:

> The Tribal cooperators' position is that the Cultural Background position is very brief, and there was no input from Bois Forte, Fond du Lac, or Grand Portage Bands as to the accuracy of the Cultural background to provide context to the evaluation. The result is that the section lacks relevant expertise and reflects little knowledge of the present-day Bands. *Again, the Tribal cooperators should have reviewed and commented on this chapter before the PDEIS was sent out.* (Minnesota Department of Natural Resources, 2009, p. 706, emphasis added)

As this comment demonstrates, the enactment of participation in environmental assessments, actions, and policies can follow legal mandates while simultaneously failing to ensure equitable participation in each case.

Indeed, this recent history regarding engagement of the Minnesota Tribal Nations in the NorthMet mine case raises concerns for the TMP. An environmental justice framework reminds us that public participation efforts can be either marginalizing or an opportunity to redress marginalizing practices. Conventionally, the legal requirement for public participation is satisfied through a combination of "notice and comment" and public meetings. In "notice and comment," agencies announce in the federal register that they are seeking public comment on a document or decision and then anyone can submit comments by mail or through a web portal like *Regulations.gov* or the BLM's *E-planning*. The agency will then read and consider all relevant comments and typically respond by adding a short section to the document at the end. For example, when the BLM renewed Twin Metals' lease in 2019, the 38,905 comments they received on the draft environmental assessment (EA)[1] resulted in the addition of two paragraphs and an appendix to the final EA. The two paragraphs indicate how many commenters participated and summarize themes: "The majority of the comments expressed opposition to renewing the federal hardrock leases. These opposing comments generally cited ecological degradation of the Boundary Waters Canoe Areas Wilderness and its watershed and socio-economic degradation as reasons for this opposition. A few comments focused on the clarity and suitability of the draft EA or the proposed stipulations. A summary of the comments and the BLM's responses to substantive comments can be found in Appendix C" (BLM, 2019, p. 11). In addition to this high-level summary, the authors provide an analysis of public comments in the appendix. According to the BLM, they performed a content analysis of comments designed to identify topics of concern and ensure all comments are fairly considered. As a final step in the process, the BLM determines whether each comment is "substantive or non-substantive in nature" (BLM, 2019, p. 70), where substantive comments do one or more of the following:

- Question the accuracy of information or analysis
- Question the adequacy of, methodology for, or assumptions used for the analysis
- Present new information relevant to the analysis
- Present a reasonable alternative other than those presented

- Questions with a reasonable basis the merits of an alternative
- Causes change in or revisions to the proposed action (BLM, 2019, p. 70)

Consistent with federal regulations, the authors respond to all "substantive" comments, which in this case included nine comments about mining impacts; nine comments about legal compliance with laws other than NEPA; 14 comments about compliance with NEPA; 17 comments about the main content of the EA; and 13 comments about the suitability of stipulations. It is worth noting that the appendix that contains the analysis of comments and subsequent responses is 35 pages, with the EA itself—the main document—totaling only 43 pages. And in this case, the comments did result in some minor changes to the EA.

In addition to public participation via notice and comment, chapter 5 of the EA features a list of tribes, individuals, organizations, or agencies consulted. The list is presented as a table with three columns: 1) name, 2) purpose and authorities for consultation and coordination, and 3) findings and conclusion. With a couple of exceptions, the name column features representatives from the area tribes and the "purpose" column is the same for each row: "The National Historic Preservation Act, The American Indian Religious Freedom Act, The Native American Graves Protection and Repatriation Act, E.O. 13007, and/or other statutes and executive orders" (BLM, 2019, p. 38). In other words, the purpose for consultation or coordination is not to seek the expertise of tribal representatives or determine their preferences—to engage meaningfully with tribal nations. Rather, the purpose is simply to comply with laws requiring basic consultation. The BLM sent letters to each representative, and then documented in "findings and conclusions" that they either didn't receive a response or that they met with the person. This approach demonstrates that stakeholders were given the opportunity to participate but seems to fall short of what a commitment to environmental justice would demand.

Conclusion

While environmental review processes are undoubtedly technical affairs, they are also infused with discourses about what matters and who counts—about what's substantive and significant. One of the critical assumptions that motivates proponents of participatory approaches is that the equity of decisions is conferred at least in part by the equity of the process (Callon et al., 2011). This brief exploration of conventional public participation practices is critical because it illustrates the tendency to rely on nondialogic mechanisms that offer an unclear relationship between participation and change. The International Association for Public Participation (IAPP) also provides some useful granularity to "participation." For the IAPP, there are five modes of participation: inform, consult, involve, collaborate, and empower. These are organized along a spectrum of public influence from *least influence* (e.g., participants are told what will be done) to *most influence* (e.g., participants are empowered and actively shape decisions and actions) (International Association for Public Participation, 2020). Broadly speaking, TC scholars have advocated for approaches that position participants in more active roles, that is, involve, collaborate, or empower.

Indeed, environmental justice emphasizes forms of participation in which participants are involved early and often, are diverse and independent, and can confirm that their contributions have been understood and considered. The resources featured in table 2.3 mirror these commitments in the context of state and federal environmental review processes, but the extent to which the DNR will enact them remains to be seen. Will the Minnesota DNR convene an advisory committee on environmental justice composed in part by tribal representatives? Will that committee be positioned as a collaborator? Will they have influence on what data is collected or ownership over what impacts are documented? In short, will the DNR prioritize the meaningful participation of those most marginalized?

Like Walton, Moore, and Jones's critical context analysis (2019, p. 143), the work I have done in this chapter seeks to understand the levels at which the problem is functioning, to help others recognize the problem, and to identify potential avenues of intervention. Toward that end, I have examined agency websites, news articles, issue-based task force reports, environmental assessments, industry and advocacy communications, and governmental policies and guidance. All of these are forms of technical communication, and as such serve in one sense as evidence of the ways these genres can instantiate marginalization and, in another sense, represent opportunities to redress inequities in environmental policy and action contexts (Walton et al., 2019; Agboka, 2018).

In addition, by examining interventional resources like *Promising Practices for EJ Methodologies in NEPA Reviews* and the efforts of the Great Lakes Fish and Wildlife Commission and other Tribal Cooperating Agencies to realize environmental justice in recent history, I hope to highlight a) ongoing efforts to bring EJ to the fore of environmental review practices and b) the importance of coalitional action surrounding the Twin Metals project and environmental issues more broadly. Advocacy efforts like the Campaign to Save the Boundary Waters Canoe Area have worked hard to amplify the concerns of those who treasure the opportunity to canoe, camp, and fish in the pristine waters north of Ely, but they should also echo the concerns of those whose treaty rights to wild rice, fish, and clean air and water are threatened, and engage meaningfully with those on the economic margins.

In addition to amplifying and extending this work in our research, we must seek opportunities to attune future technical communicators and professionals in natural resource management to injustices and the possibilities for resistance. Teaching partnerships and community-engaged learning represent one productive avenue (Butts & Jones, 2021). For example, my university offers a course called Environmental Assessment that might be enriched by a case study that examines how "social impacts" are constructed and puts multiple stakeholder perspectives, particularly Indigenous perspectives, in conversation with governmental guidance on environmental justice.

Calibrating TC commitments to advocacy with environmental justice principles enables us to ask not how many people can we include in these public participation processes, but rather how can we create the conditions in which a range of stakeholders, especially those most impacted, can meaningfully participate? How can we navigate complexity together in pursuit of just futures? Attending to the institutions, structures, and processes in which policy happens can help us identify avenues of intervention and potential collaborators. Navigating these challenges takes time and reminds us that intervention cannot be easily separated from analysis and reflection. My hope is that this chapter, in concert with those in this collection, is generative as others map their own paths toward environmental action.

Note

1. An environmental assessment may be prepared to determine whether a federal action has the potential to cause significant environmental effects. If the EA determines that impacts will be significant, a more rigorous environmental impact assessment is prepared (Environmental Protection Agency, 2022).

References

Agboka, G. Y. (2014). Decolonial methodologies: Social justice perspectives in intercultural technical communication research. *Journal of Technical Writing and Communication, 44*(3), 297–327.

Agboka, G. Y. (2018). Indigenous contexts, new questions. In A. Hass & M. Eble (Eds.), *Key theoretical frameworks: Teaching technical communication in the twenty-first century* (pp. 114–137). Utah State University Press/University Press of Colorado.

The Biden plan to secure environmental justice and equitable economic opportunity. Joe Biden for President: Official Campaign Website. (2020, October 30). https://joebiden.com/environmental-justice-plan/

Blythe, S., Grabill, J. T., & Riley, K. (2008). Action research and wicked environmental problems: Exploring appropriate roles for researchers in professional communication. *Journal of Business and Technical Communication, 22*(3), 272–298.

Bureau of Land Management. (2019). Environmental assessment. *Addition of terms and conditions for renewal of hardrock leases, MNES-001352 and MNES-001353.* https://eplanning.blm.gov/public_projects/nepa/98730/172784/209929/EA_LeaseRenewal_MNES01352-01353_FINAL.pdf

Bureau of Land Management. (2020). *BLM to prepare an environmental impact statement for proposed Twin Metals Project* [Press release]. https://www.blm.gov/press-release/blm-prepare-environmental-impact-statement-proposed-twin-metals-project

Butts, S., & Jones, M. (2021). Deep mapping for environmental communication design. *Communication Design Quarterly Review, 9*(1), 4–19.

Cagle, L. E. (2017). Becoming "forces of change": Making a case for engaged rhetoric of science, technology, engineering, and medicine. *Poroi, 12*(2), 3.

Cagle, L. E., & Herndl, C. (2019). Shades of denialism: Discovering possibilities for a more nuanced deliberation about climate change in online discussion forums. *Communication Design Quarterly Review, 7*(1), 22–39.

Callon, M., Lascoumes, P., & Barthe, Y. (2011). *Acting in an uncertain world: An essay on technical democracy.* Inside Technology.

Card, D. J. (2020). Off-target impacts: Tracing public participation in policy making for agricultural biotechnology. *Journal of Business and Technical Communication, 34*(1), 77–103.

DeVasto, D., Graham, S. S., Card, D., & Kessler, M. (2019). Interventional systems ethnography and intersecting injustices: A new approach for fostering reciprocal community engagement. *Community Literacy Journal, 14*(1), 44–64.

Eichberger, R. (2019). Maps, silence, and standing rock: Seeking a visuality for the age of environmental crisis. *Communication Design Quarterly Review, 7*(1), 9–21.

Enríquez-Loya, A., & Léon, K. (2020). Transdisciplinary rhetorical work in technical writing and composition: Environmental justice issues in California. *College English, 82*(5), 449–459.

Environmental Protection Agency. (2022). *National Environmental Policy Act review process.* https://www.epa.gov/nepa/national-environmental-policy-act-review-process

Exec. Order No. 12898, 3 C.F.R. 59 7629 (1994).

Haas, A. M., & Frost, E. A. (2017). Toward an apparent decolonial feminist rhetoric of risk. In D. Ross (Ed.), *Topic-driven environmental rhetoric* (pp. 168–186). Taylor & Francis.

Herndl, C. G., & Brown, S. C. (1996). *Green culture: Environmental rhetoric in contemporary America.* University of Wisconsin Press.

International Association for Public Participation. (2020). *Core values, ethics, spectrum: The 3 pillars of public participation.* https://www.iap2.org/page/pillars

Itchuaqiyaq, C. U., & Matheson, B. (2021). Decolonizing decoloniality: Considering the (mis) use of decolonial frameworks in TPC scholarship. *Communication Design Quarterly Review, 9*(1), 20–31.

Jones, N., McDavid, J., Derthick, K., Dowell, R., & Spyridakis, J. (2012). Plain language in environmental policy documents: An assessment of reader comprehension and perceptions. *Journal of Technical Writing and Communication, 42*(4), 331–371.

Jones, N. N., & Williams, M. F. (2017). The social justice impact of plain language: A critical approach to plain-language analysis. *IEEE Transactions on Professional Communication, 60*(4), 412–429.

Minnesota Department of Natural Resources. (2009, July). *Draft environmental impact statement (DEIS).* https://files.dnr.state.mn.us/input/environmental-review/polymet/draft_eis/volume_iii_appendices_deis_10_19_09.pdf

Minnesota Department of Natural Resources. (2020). *The Twin Metals Project: Environmental impact statement (EIS).* https://www.dnr.state.mn.us/input/environmentalreview/twinmetals/index.html

Minnesota Environmental Quality Board, & Seuffert, W., (2015). *Introducing federal national environmental policy act practitioners to the Minnesota environmental policy act process.* Environmental Quality Board.

McGreavy, B., Kelley, S., Ludden, J., Card, D., Cogbill-Seiders, E., Derk, I., . . . & Walker, K. (2020). "No(t) camping": Engaging intersections of housing, transportation, and environmental justice through critical praxis. *Review of Communication, 20*(2), 119–127.

Moore, K. R. (2017a). Experience architecture in public planning: A material, activist practice. In L. Potts & M. J. Salvo (Eds.), *Rhetoric and experience architecture* (pp. 143–165). Parlor Press.

Moore, K. R. (2017b). The technical communicator as participant, facilitator, and designer in public engagement projects. *Technical Communication, 64*(3), 237–253.

Moore, K. R., & Elliott, T. J. (2016). From participatory design to a listening infrastructure: A case of urban planning and participation. *Journal of Business and Technical Communication, 30*(1), 59–84.

Northeast Minnesotans for Wilderness. (2021). *About the campaign.* Campaign to Save the Boundary Waters. https://www.savetheboundarywaters.org/about-campaign

Robertson v. Methow Valley Citizens Council, 490 U.S. 332, 109 S. Ct. 1835, 104 L. Ed. 2d 351 (1989).

Ross, D. G. (Ed.). (2017). *Topic-driven environmental rhetoric.* Taylor & Francis.

Sackey, D. J. (2018). An environmental justice paradigm for technical communication. In A. Hass & M. Eble (Eds.), *Key theoretical frameworks: Teaching technical communication in the twenty-first century* (pp. 138–160). Utah State University Press/University Press of Colorado.

Sackey, D. J. (2020). One-size-fits-none: A heuristic for proactive value sensitive environmental design. *Technical Communication Quarterly, 29*(1), 33–48.

Simmons, W. M. (2008). *Participation and power: Civic discourse in environmental policy decisions.* State University of New York Press.

St. Amant, K. (2018). Of access, advocacy, and citizenship: A perspective for technical communicators. In G. Y. Agboka & N. Matveeva (Eds.), *Citizenship and advocacy in technical communication* (pp. xxi–xxiv). Routledge.

US Environmental Protection Agency. (2016). *Promising practices for EJ methodologies in NEPA reviews: Report of the Federal Interagency Working Group on Environmental Justice and NEPA Committee.*

Walker, G. (2010). Environmental justice, impact assessment and the politics of knowledge: The implications of assessing the social distribution of environmental outcomes. *Environmental Impact Assessment Review, 30*(5), 312–318.

Walton, R., Moore, K., & Jones, N. (2019). *Technical communication after the social justice turn: Building coalitions for action.* Routledge.

Williams, R. (1985). *Keywords: A vocabulary of culture and society* (revised edition). Oxford University Press.

Chapter 3

In Defense of a Greenspace

Students Discover Agency in the Practice of Community-Engaged Technical Communication

BOB HYLAND

Amid climate change, bee colony collapse, systemwide water supply/water quality crises, apparent increase in the occurrence and intensity of hurricanes and wildfires, and other existential environmental crises, one can be forgiven for feeling helpless. How does one rouse a sense of agency to take up impactful action against such dire threats? Those among us who choose to engage in environmental action are familiar with the powerful discursive and psychological recalcitrance adjacent to the work. In the context of technical communication (TC), this recalcitrance toward individual agency might be most felt by our students. For their entire lives, they have been surrounded by a discourse of impending doom concerning the environment. Unfortunately, the consternation about recalcitrance appears to be emerging in the cross-disciplinary, empirical literature as well.

Feelings of disempowerment and emotional strife are increasingly prevalent within the literature. Examples include documentation of the psychological impacts (Clayton et al., 2014), research on parental impact (Gaziulusoy, 2020), and the impacts of climate change communications on young people (Hibberd & Nguyen, 2013). This mounting agreement that the effects of climate change extend beyond the physical, and into the psychological, presents teachers of technical communication with both challenges and opportunities.

The challenges we face are not, per se, new challenges. For many years, TC scholarship has grappled with questions surrounding student identity formation, agency, and, more recently, advocacy. In addition, the literature gives us ample starting points to pursue continuous improvement in the use of community-engaged service-learning pedagogy as a mode for educing agency from students as they seek to participate meaningfully in action that positively impacts environmental exigencies.

What follows is a case study of how we might leverage and improve our knowledge of these assets of TC pedagogy. To amplify the process, substance, and outcomes of the case presented here, a representative review of the literature provides ground for the pedagogical approach employed throughout the 18-month, service-learning project. Following the literature review is a description of the scene in which the environmental action was focused, thereby providing the historical and geographical scope of the environmental issue at hand. The bulk of the chapter discusses artifacts of student work for environmental action, illustrating the educement of student agency and real impact in the community. Furthermore, we see empirical outcomes from the students' discovery of agency, and use of that agency in the form of technical communication for environmental action—that is, empirical outcomes such as a successful campaign to stop development in an urban greenspace, over 100 students from area schools impacted, and effective advocacy at various levels of local government and other policy arenas.

This chapter reinforces that localizing our pedagogy and coursework can empower our students to contemplate and address existential- and global-scale environmental problems such as climate change. Moreover, the case reinforces that if TC is to be positioned as a leading discipline in environmental action, then our pedagogy must include opportunities for our students to activate, and to observe the impact of, their agency in meaningful environmental action.

Our Ground: A Brief Review of Relevant Scholarship in Technical and Professional Communication

To situate both the pedagogical approach and student work in the case presented here, a representative review of the literature follows. For the purposes of grounding the case in theory, as well as to support the potential future use of this literature review as a reference, I've elected to use the following themes to broadly organize the corpus:

- community-engaged and service-learning pedagogy
- student identity
- advocacy
- agency

In a few instances, I've included literature that might not fit neatly into TC, but which can nonetheless constructively inform approaches to the instruction of technical communication for environmental action.

COMMUNITY-ENGAGED AND SERVICE-LEARNING PEDAGOGY IN TECHNICAL COMMUNICATION

The literature in our field that enunciates the benefits of a pedagogy involving students in community-engaged, service-learning coursework is robust. Youngblood and Mackiewicz (2013) develop for us a strong argument in support of the upsides to service learning in TC pedagogy and why labor toward mitigating the pedagogical limitations is worthwhile. Student-oriented benefits are also well documented, including the relationship between cognitive learning and service (Eyler & Giles, 1999; Novak et al., 2007). For these benefits to be realized, however, and for the service-learning project to reciprocally benefit the student, community partner, population served, university, and other stakeholders, the literature provides caution and guidance for assessment (Holland, 2001). Kimme Hea and Wendler Shah (2016) also illuminate the problem of hyperpragmatism in service learning, and the importance of building relationships with community partners to create more meaningful collaborations that last longer than one academic semester.

STUDENT IDENTITY FORMATION IN TECHNICAL COMMUNICATION

What the literature reveals to us about the role of student identity and positionality is also important in the context of the case presented here and, more broadly, technical communication for environmental action.

Anecdotally, the scope of an environmental exigence such as climate change (situation) may present strong forms of recalcitrance for the individual student (position) seeking effective modes of environmental action (agency). Pickering (2018) highlights the important role emotion plays in identity construction. If we figuratively cast the vast exigence of climate change as the institutional constraint a student confronts when searching for an identity in an environmental action movement, it's easy to surmise the powerful recalcitrance that students might experience when trying to locate their agency in that movement. The role of government in environmental action, and the work TC students and professionals do in or adjacent to government, would present additional challenges and opportunities related to identity, positionality, and agency in terms of discursive practices (Pickering, 2018). Our ability as TC teachers to educe agency from students and to help them see how the act of participation is a part of their identity construction (Pickering, 2019) is complicated but important work. Another important factor is the issue of authority. Lutz and Fuller (2007) elucidate the potential benefit of diminishing any presumed teacher authority and fostering a more collaborative dynamic between the teacher and students who together grapple with the agency being used for change.

ADVOCACY IN TECHNICAL COMMUNICATION

Fortunately, we have a collection specifically considering our field's attention toward its place in citizenship and advocacy (Agboka & Matveeva, 2018). The role of technical communicators is of special interest to environmental action because environmental exigencies are inextricably tied to issues of race, socioeconomics, and unequal access to participation in democratic processes resulting from testimonial and hermeneutic injustice. For instance, the health hazards of climate change have been shown to disproportionately impact underserved and marginalized populations (Ziegler et al., 2016). Furthermore, Walton et al. (2019) scrutinize the social and political structures that perpetuate disproportionate impacts of injustice and provide guidance for praxis of that theoretic. Foundational works leading up to this text provide additional informative inquiry, including the practice of TC as a humanistic endeavor (Jones et al., 2016), positioning technical communication as an ally to grassroots initiatives for empowerment (Jones, 2016), and the importance of noticing the issues that give rise to the formation of nonprofit organizations and advocacy groups in our communities in the first place (Kimme Hea & Wendler Shah, 2016). In the arena of TC instruction, particularly that which employs community engagement and service learning, our students might be well served to go beyond assignments that advocate for a specific population and to consider how their work might also impact policies at institutional or systems levels (Scott, 2004). Of course, meeting a challenge with consequences of such import as that requires a sense of agency in the student.

AGENCY IN TECHNICAL COMMUNICATION

As stated earlier, our students might most bear the brunt of a recalcitrance toward agency for environmental action as they have, for their entire lives, been exposed only to dire narratives surrounding climate— either that it is an existential crisis, or it is a lie. There is an unfortunate parallel here, I think, to Nealon's (2012) point about students only ever knowing "rampant commodification." In other words, a presumption that climate change "just is" might lead students to the resignation that they are passive critics rather than agents who can do something about it. However, our literature provides strong theoretical grounds for a pedagogy that positions TC instruction as an *educer* of agency. Wilson (2001) encourages us toward "experimentation," whereby we are "giving students permission to look beyond their preconceptions of how the system works and imagine new representations of the landscape in which they operate" (p. 95). We also encounter the notion of encouraging students to adopt a "hermeneutics of situation" that requires praxis to bend the arc toward justice, rather than a "hermeneutics of suspicion" that leaves us only to critique what is (Wilson & Wolford, 2017). Scott (2006) challenges us to take our pedagogical line further still, from one which simply exists as prescription for employment in support of existing economic systems to one that positions civic responsibility as fundamentally part and parcel to our profession. Finally, a relatively recent case provides additional agreement on increased motivation as a by-product of service learning in TC (Francis, 2018).

~

With the previous review as our ground, the case that follows traces the community-engaged, service-learning experiences of three cohorts of students enrolled in English 4093, Environmental Writing, during the Fall 2018, Spring 2019, and Fall 2019 academic semesters. Through presenting and discussing student artifacts that advocate on behalf of an urban greenspace, I argue for the efficacy of technical communication for environmental action to help students discover agency. The project initially grew out of an immediate need to defend a greenspace from a development proposal; however, through student investment in the longer-term needs of the park, as well as those of the communities and other stakeholders in direct relationship with the park, the project has evolved into an ongoing undergraduate student research and service project to protect and restore the park's ecological integrity and to address matters of equity in access to urban greenspace. Of all the positive outcomes from this ongoing community-engaged initiative, perhaps the most redeeming for many of the stakeholders is the clear indication that students who seek to engage in positive environmental action are finding the agency and platform with which to do so.

Our Scene: An "Urban Oasis" under Perpetual Threat of Development

Up a steep foothill of the Appalachian Mountains, just two miles north of where the Ohio River meets downtown Cincinnati, sit the five neighborhoods collectively known as Uptown. Within Uptown are four hospitals, a public university, two colleges, a federal research laboratory, several business districts, and countless other "brick and mortar" entities. Uptown is also flanked on its eastern and western edges by Interstates I-71 and I-75, respectively. It also contains, at the time of this writing, three of the city's seven largest employers. In other words, Uptown is a heavily built, urban environment, and there's a lot of automobile traffic.

In contrast, at the center of Uptown, where the five neighborhood boundaries converge, lie 89.3 acres of greenspace. The greenspace, named Burnet Woods and a part of the City of Cincinnati's park system, is home to, or a migrant trap for, over 160 species of birds. Burnet Woods is one of only two Audubon-designated Important Bird and Biodiversity Areas (IBA) in the county, and the smallest IBA in the United States. Unofficially referred to as an "urban oasis," one can find its visitors fishing, walking, biking, birding, picnicking, studying, or partaking in any number of other activities that seem insulated from the surrounding hubbub of urban activity outside the park. In sum, Burnet Woods is a vital greenspace for wildlife habitat and biodiversity, for ecosystem services such as mitigation of the urban heat island effect and storm water runoff, and as an outdoor respite from the stressors of urban dwelling.

Despite the beauty and ecological importance of Burnet Woods, the greenspace has been under the threat of development from its inception as a publicly managed park. In 1872, when the City of Cincinnati began to manage the land as a park, it was 170 acres in size. Ten years later, the university bought 74 of those acres for development of its campus. In 1950, the university purchased another 18 acres for additional development. Since that time, proposals to build a public library (1990), a private restaurant (2007), and a private arts center (2018) in Burnet Woods have been put before the Park Board of Commissioners. It was this last proposal, that from the arts center, that served as catalyst for creating an ongoing service-learning opportunity for students of environmental writing. The actions taken by students, an amalgamation of participatory democracy, activism, coalition building, and technical communication, were critical in the successful defense of a greenspace, and gave three separate cohorts of students the opportunity to discover agency.

Our Action: Leveraging Technical Communication and Service for Environmental Action

In the pages that follow, artifacts of student work are discussed to reveal the power of technical communication pedagogy for educement of agency. These artifacts are supplemented with narrative that describes the different situations, stakeholders, and platforms with which the students disseminated their work. Beginning in Fall 2018, the curriculum for English 4093, Environmental Writing, was designed to include community-engaged writing assignments and service learning, with the aim of supporting the resistance to development in Burnet Woods.

FALL 2018

On day one of the Fall 2018 semester, students learned that the public greenspace across the street from campus was under threat of private development. As a starting point for instruction, then, it was necessary to ask students to analyze the rhetorical situation. To do so, students were assigned the task of critically reading and writing about two documents that addressed the exigence. The first was a slide deck created by the private arts center proposing to build a 30,000-square-foot building in the park, and the second was a petition from an emerging grassroots group of stakeholders, called Preserve Burnet Woods, opposed to the development.

Figures 3.1a, 3.1b, and 3.1c are excerpts from one student's rhetorical analysis. Use of the words "backyard," "urban home," and "ecological area" indicate an understanding of the rhetorical scene. The student then juxtaposes the two competing interests in a "quiet yet critical battle." A grasp of the historical and biological importance of the scene and battle appears in the statements "has been relatively untouched for over fifty years" and "home to a wide, diverse range of species." While a seemingly straightforward opening to a rhetorical analysis, in just three sentences the student exhibits investment in the debate; there's affinity, history, and a side to take (fig. 3.1a).

As the analysis continues, the student makes the arts center an "agent" whose agency and rhetorical maneuvering can be critiqued. In developing terminology (i.e., "together," "green," "ethos," "credence," "logos," "benefits," and "logical") to describe the arts center's appeals to its audience, the student simultaneously presents the backing to understand the proposing entity's rhetorical strategy and probes the arguments advanced through that strategy (fig. 3.1b).

Similarly, in the following excerpt the student develops a terminology to describe Preserve Burnet Woods' (PBW) rhetorical approach in its petition opposing development in the park. The diction used to describe the petition, including words such as "precedent," "inciting," "conscientious," "movement," and "victimizes," suggests a budding agency in the student to assert an ethical litmus for the exigence (fig. 3.1c).

In the University of Cincinnati's backyard, a quiet yet critical battle is brewing. Burnet Woods has been relatively untouched for over fifty years and it provides an urban home to a wide, diverse range of species. While the park is an important ecological area, there are some parties interested in developing a portion of it.

Figure 3.1a. Opening sentences used by a student for rhetorical analysis of two primary sources related to the environmental exigence.

By pushing this agent, viewers will think of the arts center and the park together. The ▮▮▮▮ claim to be a grass-roots organization may not necessarily be true, but it strengthens their tie to the park by furthering a "green" image with word choice. The following slides provide the history, mission, programs, and attendance data ▮▮▮▮▮▮ up to this date, appealing to ethos by establishing themselves with the viewers, which will then give credence to their claim to the park. Appeals logos are also made by putting forward diversity data, the progress timeline, financials, and projected community and park benefits. Bringing more foot traffic and funding to the park both seem to be perfectly logical and help persuade the viewer. Ethos appeals are made with the wording of certain points, such as needing "a new home" to "nurture more inspired and insightful residents," and the use of photos of children and families enjoying the arts in the park. All of these appeals pull the viewer in and help persuade them to support the breaking ground in the park.

Figure 3.1b. Excerpt of analysis that exhibits understanding of how to deduce motive from a writer's rhetorical moves.

Credibility of PBW's stance is established by providing the history of the park, especially that the University of Cincinnati is actually sitting on the original southern half of the park, and the findings of a scientist at said university. Just as these rhetorical devices appeal to ethos, the history of little to no development in the park since the university was built and the biodiversity data help appeal to logos. Continuing the precedent of the last fifty years or so would logically help protect the park. Much of the wording appeals to pathos, inciting an emotional and personal response to the park. One important aspect of the PBW movement requires a conscientious approach to the ▮▮▮▮ need for a new location. The "envisioned structures would have served the community, but their bricks and paved parking areas would have destroyed more of the surviving parkland." This victimizes the park while also allowing PBW's image to empathize with the ▮▮▮▮.

Figure 3.1c. Additional analysis showing understanding of an evidence-based position.

Other students' rhetorical analyses of the texts exhibited similar deft. As a group, it was clear the students were wrangling with the tension between an evidence-based argument and one substantiated by pathos and ethos-as-value-proposition. This is an important point as it shows the students questioning the biases that might have informed motives behind the messaging. Further complicating the work for students was the underlying merit to the art center's claim that their building would provide opportunity for underserved youth, specifically that providing access to greenspace for underserved youth by reducing the greenspace could be characterized as misguided at best, and specious at worst.

Work on the rhetorical analysis assignment put students on strong footing to begin crafting their own arguments and to participate in the discourse surrounding the exigence. It is this educement of agency that enabled the students to move from learners in the classroom to actors in the public sphere who were effecting change on an environmental issue.

As the Fall 2018 semester progressed through October and November, various public engagement events accelerated debate over the development proposal toward the fateful December meeting at which a decision by the Park Board was to be promulgated. During this time, students began cultivating a partnership with the community group, PBW, and to fulfill communication product requests in support of PBW's community organizing. Figure 3.2 is an artifact created to satisfy one such request for the purposes of raising awareness about the pending threat to the park's ecology. Using TC as service to the community partner required careful consideration of audiences, as illustrated by the balance of quantitative and qualitative arguments in the infographic.

Pedagogically, this assignment expanded the audience for coursework to audiences beyond the classroom and challenged students to address the various stakeholders to the greenspace. The effectiveness of the infographic (fig. 3.2) and those of other students in the class increased awareness of the issue both on and off campus, including publication of a story in the student newspaper, *The News Record*—an article researched and written by a student not taking the class (fig. 3.3).

As with the infographic assignment, students adapted their writing for audiences beyond the classroom and took direct action by attending and speaking in various public forums, including Park Board meetings. For example, students wrote short speeches to accommodate strictly limited public comment periods at such forums (i.e., generally two minutes or less for each speaker). A representative artifact is included here (fig. 3.4) and

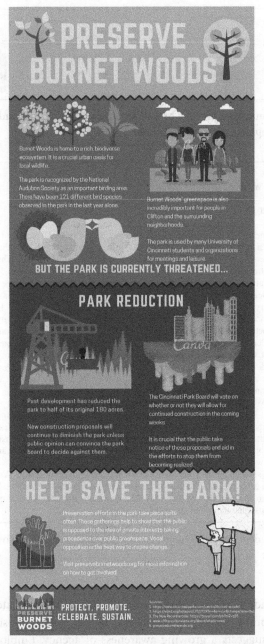

Figure 3.2. Artifact of student work in response to communication product request from our community partner.

The News Record

UC students gather to oppose Burnet Woods development

18, to protest multiple proposals to develop the historic park. Burnet Woods, an 89.3-acre greenspace operated by the Cincinnati Park Board, is ...

Oct 21, 2018

Figure 3.3. Thumbnail preview of a news article reporting on student organizing against development in Burnet Woods.

Environmental Writing

20 November 2018

Neighboring communities help each other in times of need. Here at the University of Cincinnati, our neighbor, Burnet Woods, needs our help. Burnet Woods lies adjacent to the north end of the university, which is built on the original southern half of the park. The woods serve as a safe, urban retreat for students, earning it the designation of UC's "backyard." Plenty of our student organizations, classes, and club and varsity teams utilize the greenspace for programming, learning experiences, recreation and practices. This natural oasis in the heart of the Clifton-Corryville area is enjoyed by local citizens and community groups. Here, people are able to experience nature as an escape from their urban environment—an activity proven to positively impact the mental health of city dwellers. Not only is the park a second home for many local students and citizens, it is home to hundreds, if not thousands, of species of trees, flora, and fauna. Of these organisms, many are considered rare and the migratory birds that come through the park have secured the park's categorization as an Important Bird Area, deemed so by the Audubon Society. As both UC's neighbor and an extension of our home as a backyard, Burnet Woods and all that lies within it deserves our support.

Burnet Woods has always been there for the students of UC. Now, it is time for us to be there for Burnet Woods. We must stand for its right to autonomy from outside organizations. We must stand for its protection from development. And we must stand for its species' right to not only survive, but thrive undisturbed. By raising our voices in opposition to construction, we in turn raise our voices in support of the park's preservation and protection. We can make it known that UC stands with Burnet Woods.

As both students of the University of Cincinnati and lovers of Burnet Woods, our collective voice is powerful in the face of the imposing construction plans. Join UC Students for Burnet Woods to add your dissent to the development while also adding your support for the park and all of its inhabitants and goers alike.

Figure 3.4. Student speech to recruit, organize, and mobilize opposition to development in the park.

illustrates student praxis of technical communication as oratory, as well as the adaptation of message for audience and discourse. Most importantly, the artifact suggests a growing awareness of agency. Employment of the universal "we" implies a sense of the ability to unite people. Repetition of the imperative "we must stand" suggests the student's awareness of an ability to symbolically stand, and to call others to do so with them. The conjuring power of the "collective voice" clearly reveals the student's awareness of their agency.

After weeks of work analyzing the rhetorical landscape, developing infographics to mobilize citizens, and writing and delivering oratory in public spaces, we found the fall season waning into winter, and the long semester full of service was coming to an end. However, some of the students were just getting started. They drafted a constitution, bylaws, and appropriate supporting materials to apply for, and successfully incorporate, University of Cincinnati Students for Burnet Woods (UCSBW) as a recognized student organization on campus (fig. 3.5).

This extracurricular action illustrates how student interest in the power of discourse to negotiate environmental exigencies grew throughout the semester. Ultimately, they decided to establish their own position in the discourse by establishing a student group with the mission of supporting the preservation of Burnet Woods and creating the conditions for student advocacy to continue after their own time as students would end. When looking for evidence of the educement of agency from students, none was clearer than the students' autonomous decision to activate their agency by incorporating as an official student organization. If there was any lingering doubt among those students regarding the potential impact of their agency, it was dispensed a week after the end of the semester when, on December 17, 2018, the Park Board of Commissioners voted to reject

Figure 3.5. UCSBW's student-designed logo.

the proposal for development in the park. However, the work to support our community partner, protect the greenspace from invasive species and litter, and to activate the park's latent potential as social infrastructure was just beginning.

Spring 2019

The reputation of English 4093, Environmental Writing, as a course that addresses environmental exigencies and effects change preceded the first day of class for the spring semester. On the heels of the successful campaign to stop development in Burnet Woods less than a month earlier, the new cohort of students was motivated to find out what would happen next. The community-oriented objectives of the course had pivoted. Rather than opposing an imminent threat of development, students would now need to nurture the nascent partnership with PBW, and increase service to the park and community generally, if they were to build on the hard-earned gains from the previous semester.

Before that work could begin, however, the students needed an entry point to the issue and an understanding of why the park and its surrounding communities might benefit from service by TC students. So, as before, the first assignment was rhetorical analysis of the arts center development proposal and the PBW petition. One example of student work provides two artifacts particularly useful to our discussion here. The first artifact is simply the title a student chose to use for their analysis. As you can see below (fig. 3.6), the focus of the analysis was on the formatting of the two primary sources.

While at first this may seem mundane, I find it telling about this student's perspective toward the documents. For instance, though they were no longer operatively relevant to the exigence, the student appears to find clues to each document's intended audience by way of their formatting. Furthermore, the student seems to glean the aspirational ethos of each author, that is, the arts center and PBW, respectively.

As the analysis proceeds (see fig. 3.7), the student adeptly unwinds the two rhetorical strategies and names each author's purpose; for the arts center this is to convey community "benefit," and for PBW it is to establish credibility with others so they will "support their cause." In addition, the student clearly articulates the audience each document seeks to engage. For example, the arts center "makes their argument inviting for all people," while "PBW seems to address an educated audience who have a basic

███████████████████████ and Preserve Burnet Woods (PBW) created a presentation and a petition (respectively) to promote their vision for the future of development, or lack thereof, in Burnet Woods. While both organizations emphasized different aspects of their arguments to produce a specific rhetorical strategy, the formatting of each document is indicative of significant differences in their rhetorical strategies. The differences in formatting between PBW's petition and the ███'s presentation accentuate specific aspects of their arguments and allow the organizations to tailor each argument to their respective target audiences.

One of the more distinct ways that formatting affects the rhetoric of the ███'s presentation and PBW's petition is how the organization of each document highlights specific attributes of their arguments. For example, serving the Clifton community is at the core of the ███'s argument. So, by choosing a presentation, they were able to include photos of their work in the community under the rhetorical pretense that they would continue the work if their proposal was accepted. The inclusion of pictures of their previous work illustrates the benefits the organization could continue to bring to the area without providing unnecessary written description. On the other hand, PBW chose a more formal and literary approach with their petition. This provides a professional tone (mostly from formatting but also from diction) and allows them to convey more nuanced reasons to support their cause.

The formatting of each document is also indicative of each organization's target audience. The ███'s presentation was crafted using short bullet points, easy to understand concepts, and a wide array of pictures and graphics. By structuring the presentation in this way, the ███ makes their argument inviting for all people (specifically local people because much of the content is focused on the Clifton community) regardless of education or standing. In contrast, PBW's petition was made with a single, generic photo and a long essay-like block of text, the content of which is highly specific to Burnet Woods' history and role in the environment. By focusing their argument in text with arguments centered around lofty, difficult-to-imagine concepts (i.e. "species composition," "Audubon Important Birding Area[s]," etc.), PBW seems to address an educated audience who have a basic understanding of ecosystems and their relative fragility.

Figure 3.6. Title of student's rhetorical analysis of the arts center's proposal slide deck and PBW's petition.

understanding of ecosystems and their relative fragility." These realizations, both for the students analyzing the rhetorical situation into which they would soon become agents, as well as for me as their instructor, were formative in positioning the class to leverage technical communication for environmental action.

██████████████████████ and Preserve Burnet Woods (PBW) created a presentation and a petition (respectively) to promote their vision for the future of development, or lack thereof, in Burnet Woods. While both organizations emphasized different aspects of their arguments to produce a specific rhetorical strategy, the formatting of each document is indicative of significant differences in their rhetorical strategies. The differences in formatting between PBW's petition and the ████'s presentation accentuate specific aspects of their arguments and allow the organizations to tailor each argument to their respective target audiences.

One of the more distinct ways that formatting affects the rhetoric of the ████'s presentation and PBW's petition is how the organization of each document highlights specific attributes of their arguments. For example, serving the Clifton community is at the core of the ████'s argument. So, by choosing a presentation, they were able to include photos of their work in the community under the rhetorical pretense that they would continue the work if their proposal was accepted. The inclusion of pictures of their previous work illustrates the benefits the organization could continue to bring to the area without providing unnecessary written description. On the other hand, PBW chose a more formal and literary approach with their petition. This provides a professional tone (mostly from formatting but also from diction) and allows them to convey more nuanced reasons to support their cause.

The formatting of each document is also indicative of each organization's target audience. The ████'s presentation was crafted using short bullet points, easy to understand concepts, and a wide array of pictures and graphics. By structuring the presentation in this way, the ████ makes their argument inviting for all people (specifically local people because much of the content is focused on the Clifton community) regardless of education or standing. In contrast, PBW's petition was made with a single, generic photo and a long essay-like block of text, the content of which is highly specific to Burnet Woods' history and role in the environment. By focusing their argument in text with arguments centered around lofty, difficult-to-imagine concepts (i.e. "species composition," "Audubon Important Birding Area[s]," etc.), PBW seems to address an educated audience who have a basic understanding of ecosystems and their relative fragility.

Figure 3.7. Rhetorical analysis of the arts center proposal slide deck and PBW petition completed in Spring 2019.

After the rhetorical analysis assignment provided entry into understanding the milieu of our work, it was time for the students to situate themselves in the community-engaged aspects of the coursework. To accomplish this, the soon-to-be president of PBW made a class visit. At this meeting, discussion identified needs for student service to PBW and the community at large. The prevailing need was to support the park as a source of social capital in Uptown, and to continue to build collective impact by strengthening collaboration among stakeholders. Whereas before the exigence was the immediate threat of development, the exigence had now become the perceived underutilization of the park that might lead to deficit-oriented cases for future development proposals. Fortunately, by directly interacting with our partner and being personally invited to contribute to mitigation of the problem, the students began to sense agency for creating impact.

As days became weeks and winter waned into spring, students initiated meetings with a board member of the local public school district, landed a meeting with a local school's community resources coordinator, and organized an outing in Burnet Woods for elementary school students. Throughout the process, the students used various genres of TC, including planning documents, emails, and logistical forms to facilitate bussing and activities. Furthermore, they coordinated with PBW leadership, the Parks Department, UCSBW, and the university's Ornithology Club to reserve an area of the park to provide diverse activities for the school children. The event, which ultimately took place after the end of the semester, was facilitated by most of the students from the class and a strong contingent of UCSBW and UC Ornithology Club members. Simply put, the event was a huge success. Since then, the event has become an ongoing partnership affectionately known as Nature Days, positively impacting over 100 children and adolescents from local, underserved public schools (fig. 3.8).

Students also expanded their work beyond the classroom, for example, continuing with writing a press release to announce the Nature Days event and partnerships. While some chose to apply a more reflective approach and to advance a narrative highlighting the value of service learning to themselves and the populations served, others chose a contemporary approach to simply deliver the details of the news story. In both cases, students again discovered the agency they possessed to be shapers of a narrative that advanced the park's integral role as a place for community building in Uptown. In figure 3.9, we see perhaps the strongest indication yet of a student activating agency. In describing the pedagogical roots of the service-learning course and its applied impacts, the author centers students as the agents of change. Be it the "group of students" cultivating community partnership, the "students' practice" of valuable technical communication skills, or the "student-led projects being developed": the press release makes clear to its audience that the students themselves hold the agency for environmental action.

Figure 3.8. UC and elementary school students gather in the valley of Burnet Woods park for the inaugural Nature Days.

As we can see, through use of a community-engaged curriculum and through technical communication praxis, students found in themselves the agency to impact areas in which they and the community identified the greatest need. Rather than assigning a prescriptive service plan, students were allowed to work with the community to conduct a needs assessment, to organize and mobilize the necessary human and capital resources, and to become fully vested in the objectives of the project beyond the scope of a semester-long class. While in the previous semester students leveraged praxis of technical communication for environmental action to stop development in the park, students in Spring 2019 engaged in praxis for the purposes of cultivating sustainable community partnership with

PBW, local schools, and other stakeholders, and to center the park in the discourse of community building in Uptown. In the context of the larger environmental action to defend the greenspace, this centering of the park is important for many reasons, and the centering can be clearly seen in the headline of the student press release (fig. 3.9).

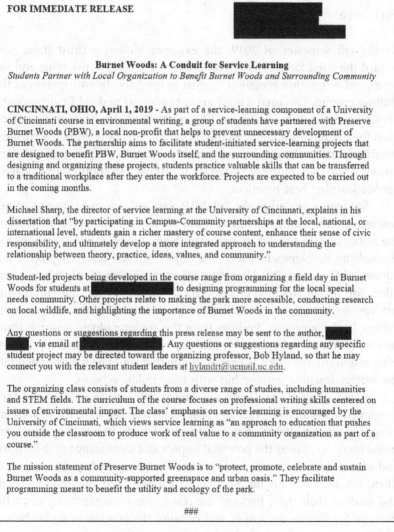

FOR IMMEDIATE RELEASE

Burnet Woods: A Conduit for Service Learning
Students Partner with Local Organization to Benefit Burnet Woods and Surrounding Community

CINCINNATI, OHIO, April 1, 2019 - As part of a service-learning component of a University of Cincinnati course in environmental writing, a group of students have partnered with Preserve Burnet Woods (PBW), a local non-profit that helps to prevent unnecessary development of Burnet Woods. The partnership aims to facilitate student-initiated service-learning projects that are designed to benefit PBW, Burnet Woods itself, and the surrounding communities. Through designing and organizing these projects, students practice valuable skills that can be transferred to a traditional workplace after they enter the workforce. Projects are expected to be carried out in the coming months.

Michael Sharp, the director of service learning at the University of Cincinnati, explains in his dissertation that "by participating in Campus-Community partnerships at the local, national, or international level, students gain a richer mastery of course content, enhance their sense of civic responsibility, and ultimately develop a more integrated approach to understanding the relationship between theory, practice, ideas, values, and community."

Student-led projects being developed in the course range from organizing a field day in Burnet Woods for students at ▮▮▮▮▮▮▮▮▮▮ to designing programming for the local special needs community. Other projects relate to making the park more accessible, conducting research on local wildlife, and highlighting the importance of Burnet Woods in the community.

Any questions or suggestions regarding this press release may be sent to the author, ▮▮▮▮ , via email at ▮▮▮▮▮▮▮▮▮▮▮ . Any questions or suggestions regarding any specific student project may be directed toward the organizing professor, Bob Hyland, so that he may connect you with the relevant student leaders at hylandrt@ucmail.uc.edu.

The organizing class consists of students from a diverse range of studies, including humanities and STEM fields. The curriculum of the course focuses on professional writing skills centered on issues of environmental impact. The class' emphasis on service learning is encouraged by the University of Cincinnati, which views service learning as "an approach to education that pushes you outside the classroom to produce work of real value to a community organization as part of a course."

The mission statement of Preserve Burnet Woods is to "protect, promote, celebrate and sustain Burnet Woods as a community-supported greenspace and urban oasis." They facilitate programming meant to benefit the utility and ecology of the park.

###

Figure 3.9. Press release emphasizing outward benefits of service-learning course-work.

By skillfully and energetically completing their coursework, by further developing the UCSBW student group as a voice in the discourse surrounding Burnet Woods, by deepening the partnership with PBW, and by making measurable impact through their service work, students in the Spring 2019 Environmental Writing class discovered and leveraged the agency available to them.

FALL 2019

By the fall semester of 2019, the exigence shifted a third time, now toward the need for greater appreciation of the ecosystem's value and the ecological restoration measures needed in the park. In anticipation of the shifting ground, I secured a mini-grant to support making Environmental Writing a research-based course. The course had also become a capstone class for the Environmental Studies program, therefore adding a stronger interdisciplinary element to assignment objectives. What didn't change, however, was the rooting of pedagogical approach in community-engaged, service-learning best practices.

Students enrolled in the fall 2019 class self-selected into one of five research teams: the microclimate team, the community perceptions team, the storm and wastewater management team, the bird diversity team, and the website development team. The object of each research project was connected to an associated benefit, such as mitigation of urban heat island affect for the microclimate study, equity in access for the community perceptions investigation, and habitat preservation for the bird diversity survey. In the aggregate, findings from this undergraduate research could be used to inform policy, to affect the discourse away from anthropocentric bias, and to inform STEAM programming that could be incorporated into the service-learning partnership with local public schools.

Once formed, each team conducted a literature review, and then developed a research abstract and proposal. From the start, the educement of agency was enabled by showing students how to situate their own work in the literature, assess the potential impact and significance of their work, and advocate for the resources they would need to conduct their research. Then, the race was on for students to execute their method, aggregate and analyze their data, package the results for dissemination, adapt the message for various audiences, and enhance the programming for Nature Days in just one semester's time.

Two artifacts of the science/issue brief assignment illustrate the pedagogy's educement of student agency. We can see this agency by looking at their strategic choices on how to leverage empirical results from research for the purposes of engaging different audiences. One artifact, a science brief (figs. 3.10a and 3.10b), conveys well the student's comprehension of the conventions of technical communication. Opening with a statement of impact, the student makes clear the effect of her team's research on "policy

---

Tree Canopies Mitigating Urban Heat Island Effect

IMPACT STATEMENT

Understanding the functions of a tree canopy and its effects on microclimate within an urban environment is critical for mitigating urban heat island effect where increasing temperatures pose a risk towards maintaining a healthy and sustainable urban environment. As urbanization continues to accelerate, a study conducted in Cincinnati indicated rising temperatures and changes in air quality are creating health risks for the urban population, specifically marginalized communities (Beck et al., 2016). However, tree canopies can mitigate many of these effects. Studies, such as this one, are vital if we are to obtain scientific understanding and policy implementation in support of preserving and expanding tree canopies in urban environments.

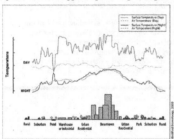

Figure 1. Diagram of Urban Heat Island Effect (U.S. EPA, 2008)

BACKGROUND

As an Environmental Studies Capstone project team at the University of Cincinnati, our investigation is part of the larger, interdisciplinary New Green Uptown project which is looking at UC/Community relations through the lens of greenspace assessment around the UC campus. We proposed, developed, and are investigating the microclimate in Burnet Woods park. Furthermore, the sharing of our methods and results with the Planning and Geography courses involved in the interdisciplinary series may help inform student work in those courses by revealing the potential microclimate effect of greenspaces in neighborhoods being studied by those students.

In the broader context, microclimate studies in urban greenspaces are important because greenspaces are often the most effective mediators of heat island effect in urban environments. With global temperatures expected to soon average at least 1.5°C above preindustrial levels, advancing our understanding of warming mitigation by urban greenspaces is critical (IPCC, 2018). If it can be shown that greenspaces with mature tree canopy can significantly reduce peak day time temperatures, buffer built environments from extreme wind events, and provide other microclimate mitigation service, these results may help to inform public policy that promotes greenspace protection and expansion to moderate the negative effects of climatic warming.

PROJECT DESCRIPTION

One ENGL and five EVST undergraduates at the University of Cincinnati as part of an interdisciplinary course series are measuring temperature, humidity, dewpoint, windspeed, evaporation rates, and soil conditions under a tree canopy and in an open area in Burnet Woods. Data will be collected during peak points of the day (i.e. sunrise, mid-afternoon, and sunset) and will undergo comparative analysis as well. Temperature, humidity, dewpoint, and windspeed will be collected via weather station for approximately ten minutes during the peak points of the day. Evaporation rates will be measured by placing water in aluminum pans under the tree canopy and in an open area, waiting an hour, and measuring the amount left over from evaporation. This quantifiable data will be collected and analyzed through data tables and graphs to visualize these results.

Figure 3.10a. Page 1 of a science brief for the microclimate investigation.

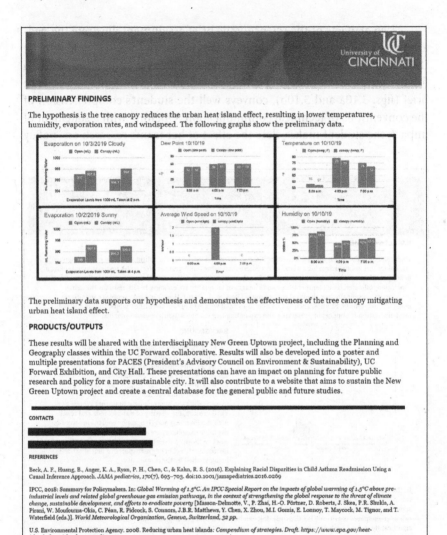

Figure 3.10b. Page 2 of a science brief for the microclimate investigation.

implementation in support of preserving and expanding tree canopies." Moreover, she summarizes the background, method, preliminary results, and outcomes in accordance with conventions of the genre.

Where the science brief reported on preliminary findings from the microclimate investigation, an issue brief revealed initial findings from the community perceptions survey (figs. 3.11a and 3.11b). This genre shift

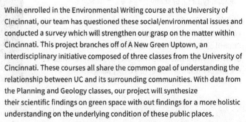

Community Perception Towards Green Space

Impact Statement:

Acknowledging the perceptions and attitudes which individuals have towards green space within their community is essential for a more in-depth analysis of how green space affects their lives, whether positively or negatively. As cities further develop and consume natural areas, marginalized groups that have the least amount of control over such issues are often affected the most. With this, access to quality green space has changed into a matter of social equality. Overall, understanding community perception towards green space can help illuminate the unique barriers within varying neighborhoods and how this contributes to citizen's ability to reap the physical and mental benefits that green space has to offer.

Background:

While enrolled in the Environmental Writing course at the University of Cincinnati, our team has questioned these social/environmental issues and conducted a survey which will strengthen our grasp on the matter within Cincinnati. This project branches off of A New Green Uptown, an interdisciplinary initiative composed of three classes from the University of Cincinnati. These courses all share the common goal of understanding the relationship between UC and its surrounding communities. With data from the Planning and Geology classes, our project will synthesize their scientific findings on green space with out findings for a more holistic understanding on the underlying condition of these public places.

The New Green Uptown initiative, along with this project, in particular, is crucial for the future of Cincinnati. Research has shown that there is a growing issue, emerging from a rise in quantity over the quality of green space. The notion that green space provides valuable benefits for urban dwellers such as lower stress levels, enhanced productivity, and decreased asthma reports are understood. However, these benefits are a privilege, making green space a socio-economic issue, were communities that would benefit from these attributes the most experience green space the least. (Landscape and Urban Planning, 2018) By conducting these surveys, our group will understand the perceptions and unique barriers that community members experience with green space, which can later apply to solutions for more equitable distribution and regulation.

Figure 3.11a. Page 1 of an issue brief for the community perceptions survey.

shows both intuition and agency from the student to select the kind of TC that would best serve her research team's audience, and best advance her purpose in writing the document.

Project Description:

This project is composed of six students from the Environmental Writing course, a mixture of EVST, ENGL, and ORGN majors. Derived from A New Green Uptown, our focus is on community perceptions and attitudes towards green space, including the barriers and benefits that people see within their communities. Throughout the 2019 Fall Semester, data collection will be obtained and then analyzed in the five communities surrounding Burnet Woods: Avondale, Mt. Auburn, Clifton, Cuff, and Corryville. We will achieve this data through the distribution of surveys that contain both qualitative and quantitative inquiries. These findings will then we entered into an excel sheet for organizational purposes. With these results, we will compare and contrast the responses seen between the different neighborhoods, and apply our findings to UC's further developing plans, illuminating the risks of further encroachment.

Preliminary Findings :

In regards to the expected responses of this survey, we hypothesize that the answers will vary between the neighborhoods for numerous contributing factors. As of now, our preliminary findings elicit many of these speculations. The overwhelming majority of our surveys find that community members value green space for its effect on health, including better air quality and a sense of tranquility. However, neighborhoods also communicate a disconnect with green space for a multitude of reasons such as quality, time, or safety.

Outputs :

Future outputs of this project will be presented to City Council on November 19th, compiled within The New Green Uptown's emerging webpage, and also presented at the end of the fall semester to a larger audience.

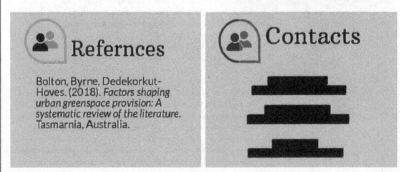

Refernces

Bolton, Byrne, Dedekorkut-Hoves. (2018). *Factors shaping urban greenspace provision: A systematic review of the literature.* Tasmarnia, Australia.

Contacts

Figure 3.11b. Page 2 of an issue brief for the community perceptions survey.

The student clarifies two important matters in the statement of impact. The first is that, without a broad-based, community-engaged understanding of the perceived value of greenspace in the community, no policy for the protection or management of greenspace is likely to enjoy sustained effectiveness. Second, she makes clear that marginalized groups, who often face testimonial and procedural injustices in democratic processes generally, are also likely marginalized from the discourse about greenspace and therefore disproportionately impacted by mismanagement of those spaces. She states, "Marginalized groups that have the least amount of control over such issues are often affected the most."

In the preliminary findings section, the student cites the substantial evidence of agreement that greenspace is important to the community broadly speaking, but that different groups value the greenspace for different reasons. Here, again, we see student agency. By way of her position as student researcher and technical communicator, she makes statements of impact and significance that are substantiated by empirical evidence. Furthermore, she feels empowered to advocate for marginalized stakeholders in the study population. Although she might have elected to develop a science brief as others did (figs 3.12a and 3.12b) and present just the quantitative aspects of her research, the student instead amplified the voice of people who are marginalized in the form of an issue brief.

Finally, I arranged for the student research teams to present their preliminary findings at a meeting of the Education, Innovation, and Growth Committee of Cincinnati's City Council. Beyond the obvious experiential-learning benefits of delivering a formal presentation at City Hall, on camera, and in front of city officials, the assignment also allowed for discussions about how to translate scientific and technical information for policymakers. Moreover, it provided students with the opportunity to locate agency in the use of technical communication to advocate for a specific stakeholder: the public, including its underrepresented segments.

Importantly, Burnet Woods is a public park, and yet proposals for development in the park come from private entities. To center the public voice in negotiations on how to manage a public park, the website development team created a clearinghouse for interdisciplinary, community-based research. The subhead students used on the title slide (fig. 3.12a) makes clear their aim of "facilitating interdisciplinary collaboration and community engagement." To be presenting within the epicenter of public affairs (i.e., City Hall), and to name a cornerstone of participatory democracy (i.e., "community engagement") for the audience of elected officials, illustrates that the students were leveraging their agency as technical communicators.

On the second slide (fig. 3.12b), we see that "community engagement" and, more specifically, "procedural justice" is central to their work of building a digital space for public empowerment. The students proceed

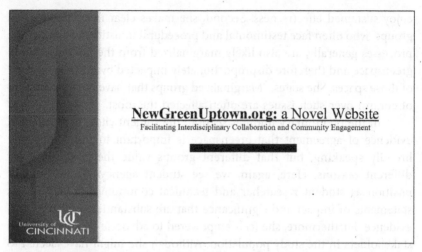

Figure 3.12a. Cover slide for the website development team presentation.

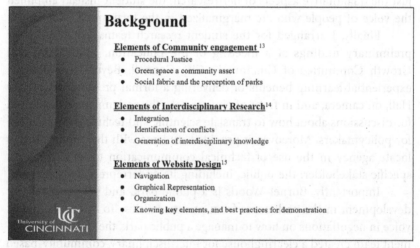

Figure 3.12b. Web development team grounding their project in the literature.

to frame greenspace as a "community asset" and part of the "social fabric." By developing a clear summary of the complex relationships between research, engagement, and management of public lands, these students exhibit how technical communication praxis not only educed agency for them, but also how they used that newfound agency to serve and facilitate agency for their stakeholders.

Slide three (fig. 3.12c) continues this advocacy for the public's ability to participate in greenspace management, as we see the web design priorities are aimed at "spreading information," "engaging community members," and "facilitating communication." Similarly, slide four in the student presentation (fig. 3.12d) represents the site architecture, which invites users to "browse research projects," "stay up to date with findings," and "get involved in the community." On the projection screens of the

Figure 3.12c. Web development team presenting community benefits of their project.

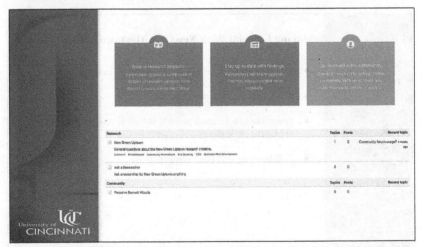

Figure 3.12d. Web development team presenting technical structure of their project.

City of Cincinnati's council chambers, undergraduates presented details on their website being designed to empower citizens to take action toward stewardship of public greenspace. The irony of students doing the job of what is arguably the city's job (i.e., effective community engagement) was not lost on anyone in chambers that day, and their presentation serves as another example of technical communication pedagogy as an educer of agency for environmental action.

On their final slide (fig. 3.12e) and in their spoken word that accompanied the slide, the web development team made apparent to City Council the class's intention to influence policy going forward through "sharing," "sustaining," "promoting," and "collaborating."

In concert with the other four student presentations that day, this presentation raised eyebrows by revealing the gap in how citizens are provided access to decision making about public spaces. That day the five research teams leveraged the power of technical communication for environmental action by developing evidenced-based visual and spoken arguments (agency) in defense of the greenspace (advocacy) and the stakeholders who are marginalized from processes that affect management of that space (discourse).

After two and a half months of community-engaged research and service, students put a lid on data collection and analysis and prepared poster presentations for the semester-end exhibition. I invited city leadership and staff; members of the neighborhood councils for Avondale, Clifton,

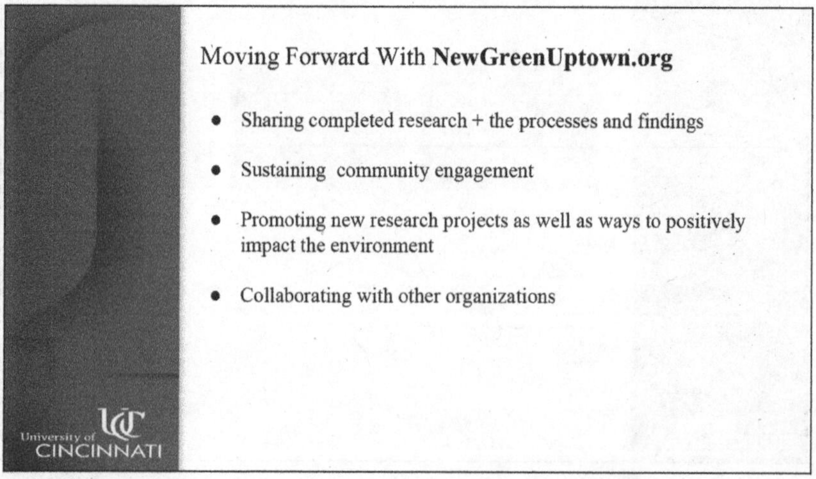

Moving Forward With **NewGreenUptown.org**

- Sharing completed research + the processes and findings

- Sustaining community engagement

- Promoting new research projects as well as ways to positively impact the environment

- Collaborating with other organizations

University of CINCINNATI

Figure 3.12e. Web development team sharing vision and calling for collaboration in next steps.

Corryville, CUF, and Mt. Auburn; our community partner, PBW; other stakeholders to the park; and the public to a two-hour presentation and celebration of the students' work. Equipped with the lessons of attending and speaking at neighborhood councils, PBW events, City Hall, and many other informal settings, and flanked by their scientific posters, the students accessed their budding expertise and delivered interdisciplinary, empowered advocacy in defense of the greenspace.

In hindsight, it's difficult to pin down what might be the foremost impact of the first 18 months of this ongoing community-engaged TC course. The successful campaign to stop development in the park gave way to the Spring 2019 cohort cultivating community partnerships, and to the Fall 2019 cohort spearheading research- and communication-based action for better policy. If we circle back, however, to the recalcitrance our students experience when seeking agency in environmental activism, there may be no better evidence of impact than those same students using their newfound agency to educe similar agency from students even younger still. While conducting research, negotiating multiple and sometimes competing discourses in the community, and presenting findings to various stakeholders, the Fall 2019 class continued and expanded the service-learning partnership with the local elementary school. Programming extended to classroom visits at the school and science-based learning activities in the park (fig. 3.13). So, maybe, the biggest impact will manifest years from

Figure 3.13. UC students facilitate a water sample preparation activity at Nature Days in Burnet Woods.

now when some of the elementary school students enroll at university and sign up to be leaders of UCSBW, or a group like it, with a sense of agency already in hand.

Implications

Helping students bridge the theory and principles of technical communication to praxis in an applied setting has, for many years, been the aim for much of our pedagogical work. As is shown in the Our Ground section of this chapter, we need not look far for insightful literature that can inform how we might incorporate community engagement, service learning, and advocacy into our instructional practices. In doing so, however, we might consider further prioritizing for students the principled role of civic responsibility in TC, a prioritization that this chapter shows as having the capacity to educe agency from students.

This might be particularly true when it comes to the praxis of technical communication for environmental action. By virtue of their scope and complexity, the environmental exigencies with which our students grapple are of a kind that all but guarantee recalcitrance of several kinds. And so, by clarifying that TC praxis is more than simply a means to an end for environmental action, we might also show that it is a responsibility that calls for civic dutifulness, advocacy, and a willingness to cross discursive boundaries. In turn, we may also show students that answering this call to responsibility creates agency.

In the case presented here, the initiative by students to form UCSBW and solidify their place in the discourse surrounding Burnet Woods is but one example of the relationship between responsibility and agency. The charter cohort of UCSBW felt a responsibility to the greenspace, the community, and future students at the university and, therefore, established a collective student voice in the discursive arena. Through that responsibility they found the agency to incorporate a student organization, cultivate partnerships on and off campus, and advocate for marginalized stakeholders. Other examples presented in this chapter further suggest the important relationship between a sense of civic responsibility and the self-detection of agency.

To put it another way, I'm arguing the case presented here provides evidence that, if we only frame for our students that technical communication *can* be used to make a difference, we may be missing an opportunity for the educement of agency. Instead, then, I'm positing that we provide opportunities for our students to see that technical communication *must* be used to make a difference—there is an inherent responsibility we have as technical communicators to undertake environmental action, and to serve those disproportionately impacted by environmental injustice. As shown in this chapter, it appears that accepting this responsibility led to agency, and activating that agency resulted in real impact ranging from defeating a development proposal to effectively cultivating a sustained community partnership, to leveraging research to affect policy at the local level.

This kind of pedagogical work is complicated and can be "messy" as we say. Our responsibility, then, is to continue the research of TC pedagogy in a manner that ensures best possible outcomes for our students and all other stakeholders in our curricular activities. In accepting this responsibility, we, too, might discover latent agency that will help us prepare our students to be agents in the praxis of technical communication for environmental action. And, as is increasingly clear as climate change accelerates, the environment needs technical communicators now more than ever before.

Permissions

All artifacts of student work in this chapter are included with obtained written permission from the student authors responsible for those artifacts. A Non-Human Subjects Research Determination was obtained from the University of Cincinnati's Institutional Review Board. IRB ID: 2020-0530.

References

Agboka, G. Y., & Matveeva, N. (2018). *Citizenship and advocacy in technical communication: Scholarly and pedagogical perspectives.* Taylor & Francis. doi:10.4324/9780203711422

Clayton, S., Manning, C. M., & Hodge, C. (2014). *Beyond storms and droughts: The psychological impacts of climate change.* American Psychological Association and ecoAmerica.

Eyler, J., & Giles, D. E., Jr. (1999). *Where's the learning in service-learning?* Jossey-Bass Higher and Adult Education series. Jossey-Bass.

Francis, A. M. (2018). Community-engaged learning in online technical communication classes: A tool for student success. In G. Y. Agboka & N.

Matveeva (Eds.), *Citizenship and advocacy in technical communication: Scholarly and pedagogical perspectives* (1st ed., pp. 223–242). Routledge. doi:10.4324/9780203711422-11

Gaziulusoy, A. İ. (2020). The experiences of parents raising children in times of climate change: Towards a caring research agenda. *Current Research in Environmental Sustainability, 2.* doi:10.1016/j.crsust.2020.100017

Hibberd, M., & Nguyen, A. (2013). Climate change communications and young people in the kingdom: A reception study. *International Journal of Media and Cultural Politics, 9*(1), 27–46. doi:10.1386/macp.9.1.27_1

Holland, B. A. (2001). A comprehensive model for assessing service-learning and community-university partnerships. *New Directions for Higher Education, 2001*(114), 51–60. doi:10.1002/he.13

Jones, N. N. (2016). The technical communicator as advocate: Integrating a social justice approach in technical communication. *Journal of Technical Writing and Communication, 46*(3), 342–361. doi:10.1177/0047281616639472

Jones, N. N., Moore, K. R., & Walton, R. (2016). Disrupting the past to disrupt the future: An antenarrative of technical communication. *Technical Communication Quarterly, 25*(4), 211–229. doi:10.1080/10572252.2016.1224655

Kimme Hea, A. C., & Wendler Shah, R. (2016). Silent partners: Developing a critical understanding of community partners in technical communication service-learning pedagogies. *Technical Communication Quarterly, 25*(1), 48–66. doi:10.1080/10572252.2016.1113727

Lutz, J., & Fuller, M. (2007). Exploring authority: A case study of a composition and a professional writing classroom. *Technical Communication Quarterly, 16*(2), 201–232. doi:10.1080/10572250709336560

Novak, J. M., Markey, V., & Allen, M. (2007). Evaluating cognitive outcomes of service learning in higher education: A meta-analysis. *Communication Research Reports, 24*(2), 149–157. doi:10.1080/08824090701304881

Nealon, J. (2012). *Post-postmodernism: Or, the cultural logic of just-in-time capitalism.* Stanford University Press.

Pickering, K. (2018). Navigating discourses of power through relationships: A professional and technical communication intern negotiates a meaningful identity within a state legislature. *Journal of Technical Writing and Communication, 48*(4), 441–470. doi:10.1177/0047281617732019

Pickering, K. (2019). Emotion, social action, and agency: A case study of an intercultural, technical communication intern. *Technical Communication Quarterly, 28*(3), 238–253. doi:10.1080/10572252.2019.1571244

Scott, J. B. (2004). Rearticulating civic engagement through cultural studies and service-learning. *Technical Communication Quarterly, 13*(3), 289–306. doi:10.1207/s15427625tcq1303_4

Scott, T. (2006). Writing work, technology, and pedagogy in the era of late capitalism. *Computers and Composition, 23*(2), 228–243. doi:10.1016/j.compcom.2005.08.008

Walton, R., Moore, K. R., & Jones, N. N. (2019). *Technical communication after the social justice turn: Building coalitions for action* (1st ed.). Routledge. doi:10.4324/9780429198748

Wilson, G. (2001). Technical communication and late capitalism: Considering a postmodern technical communication pedagogy. *Journal of Business and Technical Communication, 15*(1), 72–99. doi:10.1177/105065190101500104

Wilson, G., & Wolford, R. (2017). The technical communicator as (post-postmodern) discourse worker. *Journal of Business and Technical Communication, 31*(1), 3–29. doi:10.1177/1050651916667531

Youngblood, S. A., & Mackiewicz, J. (2013). Lessons in service learning: Developing the service-learning opportunities in technical communication (SLOT-C) database. *Technical Communication Quarterly, 22*(3), 260–283.

Ziegler, C., Morelli, V., & Fawibe, O. (2016). Climate change and underserved communities. *Primary Care, 44*(1), 171–184. doi:10.1016/j.pop.2016.09.017

Chapter 4

Flood Insurance Rate Maps as Communicative Sites of Pragmatic Environmental Action

Daniel P. Richards

Let's begin by getting the obvious out of the way first: flood insurance is a banal, if not strange, topic to cover in a collection on environmental action. It is certainly not the topic I anticipated myself working with a couple years ago, even as I was writing and researching about sea level rise communication in my coastal community. And it even seems quite far afield from technical communication (TC) and its typical disciplinary inquiries—even beyond the scope of problems we could solve. What I've learned over the past two years in being part of a state-supported grant project with the aim of increasing the number of flood insurance policies held by residents in our community is that in fact that is not true: despite its bland affectations, flood insurance is an engaging area of inquiry that merges research in cartographical rhetorics, environmental plain language, zoning regulations and environmental justice, risk literacy and community engagement, and global climate change discourse and therefore presents an array of environmental problems technical communicators can help solve. Flood insurance is an area of inquiry that requires a fair amount of traditional technical communication work, especially in "simplifying" the complex intersection of environmental and actuarial information.

This chapter recounts just such an effort and in doing so frames flood insurance enrollment rates as an environmental problem to be solved under our field's purview.

As someone who didn't own a home at the beginning of this project, I knew next to nothing about flood insurance. I assumed, like many others, that flood insurance was covered by homeowners' insurance in minor cases and that FEMA pays homeowners a fair bit in major cases. Neither is true. I also assumed that homeowners and businesses residing in high and medium flood risk zones were required to have flood insurance. Not always true. And I also assumed that flood insurance rate maps (FIRMs) were highly accurate in anticipating where the water goes once storm surge comes ashore. The answer? Quite untrue. Fortunately for me, research in the fields of public policy and public understanding of science alleviates my concerns that I was the last one to find these assumptions to be false (Dixon et al., 2006; Kousky, 2016; Michel-Kerjan et al., 2015). Time and again we see the American public—even those in high-risk regions—being misinformed about or improperly calibrated to the risks and realities of flooding, save for those who had already endured a catastrophic storm event or property damage (Shao et al., 2017). As with most of our risk analyses and assessments, experience proves to be the most poignant and prolific of persuaders.

So then here is the problem posed for communities and regions who have yet to experience the persuasive education provided by Hurricanes Harvey, Maria, Sandy, or Michael: How do you increase the number of residents and business owners holding active flood insurance policies *before* the big one comes? The number of flood insurance policies an affected region holds is a significant factor in the community's resilience and ability to rebuild itself, but after each major hurricane or tropical storm, post-disaster analysis reveals that only a fraction of homes and businesses destroyed or damaged possessed a flood insurance policy. For Harvey in Houston, that number was around 18% (Pralle, 2017). Continued expansion and development of major cities in concert with climate change means that the frequency of storms and their meandering pathways are increasingly difficult to predict. Flood maps and their articulated risk zones need constant updating and terminology needs continual refinement, especially as Houston recently faced its third 500-year flood in the past three years (Ingraham, 2017). How do regional authorities work toward community resilience in rebuilding when the maps projecting flood risk are increasingly less accurate and the terminology used to communicate them to the public unclear?

This chapter discusses but one solution to this massive problem: Get as many residents as possible living in low- to medium-risk flood zones to obtain a flood insurance policy to increase community resilience. This is not simple, and there are many ways to accomplish this that all need to work together simultaneously. Our way was to develop a calculator that would help residents learn how much a flood insurance policy would cost in terms of a yearly premium. The assumption of the grant project was that one of the main reasons why participation in the National Flood Insurance Program (NFIP) is so low is because there are misconceptions about cost. If only there were a tool that could calculate yearly premiums without even talking to an insurance company or the government first, and the user could see that the yearly premium was affordable, the community would become more resilient. I write this chapter not with any sort of expertise on flood insurance and policy but as someone currently exploring the possibilities for the field of technical communication of further engaging with flood insurance and its documentation and political ecology. I also write this chapter as a UX "team of one" (Buley, 2013), having just developed a flood rate calculator from scratch based, quite literally, on the obscure appendix J of the NFIP rate tables—a calculator that was an effort in plain language and in technical procedurality. I argue that insuring our currently built communities against inevitable disaster is a form of environmental action and therefore an increase in flood insurance policies held by residents is a solution, and a measurable one at that. This chapter makes a case for more work to be done by technical communicators in the area of flood insurance, but it also asks the reader to consider the value of more procedural, banal work that can be done now as it relates to environmental resilience in flood-prone communities.

The Dialogic Nature of Flood Insurance Rate Maps

One of the neat things about flood insurance rate maps (FIRMs) is that they are the *actual* actuarial framework of flood insurance, and the complex rate table (appendix J, see fig. 4.1) is the supplement. So often in the field of risk communication the visual itself is augmentation, a supplement to help interpret a complex table. In this case the map dictates flood zone

RATE TABLE 2A. REGULAR PROGRAM – PRE-FIRM CONSTRUCTION RATES[1,2,3]

ANNUAL RATES PER $100 OF COVERAGE (Basic/Additional)

FIRM ZONES A, AE, A1–A30, AO, AH, D[4]

	OCCUPANCY	SINGLE FAMILY		2-4 FAMILY		OTHER RESIDENTIAL		NON-RESIDENTIAL BUSINESS[5]		OTHER NON-RESIDENTIAL[5]	
		Building	Contents	Building	Contents	Building	Contents	Building	Contents	Building	Contents
BUILDING TYPE	No Basement/Enclosure	1.27/1.17	1.60/2.08	1.27/1.17		1.27/2.45		3.60/6.76		1.38/2.55	
	With Basement	1.36/1.71	1.60/1.76	1.36/1.71		1.27/2.04		3.79/6.60		1.46/2.51	
	With Enclosure[6]	1.36/2.05	1.60/2.08	1.36/2.05		1.36/2.53		3.79/8.35		1.46/3.15	
	Elevated on Crawlspace	1.27/1.17	1.60/2.08	1.27/1.17		1.27/2.45		3.60/6.76		1.38/2.55	
	Non-Elevated with Subgrade Crawlspace	1.27/1.17	1.60/1.76	1.27/1.17		1.27/2.45		3.60/6.76		1.38/2.55	
	Manufactured (Mobile) Home[7]	1.27/1.17	1.60/2.08	1.27/1.17				3.60/6.76		1.38/2.55	
CONTENTS LOCATION	Basement & Above[8]				1.60/1.76		1.60/1.76		7.15/11.33		2.70/4.27
	Enclosure & Above				1.60/2.08		1.60/2.08		7.15/13.60		2.70/5.10
	Lowest Floor Only – Above Ground Level				1.60/2.08		1.60/2.08		7.15/5.93		2.70/2.25
	Lowest Floor Above Ground Level and Higher Floors				1.60/1.46		1.60/1.46		7.15/5.06		2.70/1.94
	Above Ground Level – More Than 1 Full Floor				.35/.12		.35/.12		.24/.12		.24/.12
	Manufactured (Mobile) Home[7]								7.15/5.93		2.70/2.25

Figure 4.1. National Flood Insurance Program's specific rating guidelines (appendix J). Fair use.

designations, and once users locate themselves on the FIRM (fig. 4.2), they can begin the process of better understanding what their risk is and what the yearly policy premium might be. Naturally, however, FIRMs change. Flood zone boundaries are extended or brought in. Amendments are made. Politicians, city planners, and wealthy coastal residents challenge designations (Dedman, 2014). Changes brought by development, other construction projects, and sea level rise are accounted for. The fields of technical communication and environmental rhetoric have overlooked these flood maps as a topic even though these maps operate as the authoritative artifacts delineating flood hazard areas across the United States and even though flooding is the nation's leading cause of deaths and property damage by disaster. More than just serving as the actuarial framework of the National Flood Insurance Program (NFIP), flood maps are also used for guidance in disaster mitigation, land use planning, and emergency response as well as, and perhaps most pertinently for readers of this collection, the communication of risk to the public.

Figure 4.2. Screenshot of flood insurance rate map in Norfolk, Virginia.

Given the fluidity of these maps, the theoretical framework for engagement is already in place. The FIRMs themselves are seen by their own designers, developers, and users as modifiable, an ever-changing depiction of potential risk based on uncertain data projections. In fact, Mark Monmonier's *Cartographies of Danger: Mapping Hazards in America* (1997), the definitive cartographical analysis of the origin and development of flood insurance rate maps, characterizes FIRMs as inherently dialogic:

> As a fixed, detailed image of an elusive abstraction (risk of flooding), the Flood Insurance Rate Map is like a historical novel: an easily grasped fictionalized representation that helps readers understand (or think they understand) a complex phenomenon. However politically and societally useful, flood-zone boundaries are not real—as moving targets, they can't be. But by describing risk as plausibly greater here than there, these cartographic caricatures provide a geographic foundation essential for the actuarial framework that makes flood insurance possible. And as part of a larger process (loss reduction), flood maps *initiate a dialogue* for limiting damage in flood prone areas. (p. 126, emphasis added)

Unlike the work of scholars in technical communication to investigate the *unapparent* ideological (Barton & Barton, 1993), argumentative (Propen, 2007), and political (Welhausen, 2015) properties of maps, Monmonier's monograph need not peel back any invisible layering: the very historical purpose of flood maps was dialogic *by design*. The objective of the FEMA-administered National Flood Insurance Program (NFIP) as it was established in 1968 was to "identify and map flood prone communities and to make flood insurance available in communities that adopt and enforce floodplain management regulations (e.g., zoning, building requirements, special-purpose floodplain ordinances)" (NRC, 2009, p. 13). If a community wishes to have its residents eligible for nationally subsidized flood insurance, then it will enforce the regulations. The maps are an agreement of perception and action between federal agencies and local communities. (Note that over 21,000 communities currently participate.) Should a resident or business owner receive a certificate with contrary evidence, they can file a letter of map amendment (LOMA) and literally change the zoning of the map. The maps are a cartographical conversation between scientists, engineers, and residents. They are designed to be changed.

And yet, despite these dialogic qualities, a persistent challenge to the NFIP remains: getting residents who live in moderate or minimal risk flood zones to purchase a flood insurance policy. Homeowners and business owners in the highest risk areas (special flood hazard areas, or SFHA) are required to purchase a flood insurance policy if the mortgage is federally regulated or insured (most are). These areas have been assessed to be at the highest risk of flooding and persons residing or operating in these zones have no choice. Residents in "moderate" or "minimal" risk flood zones are encouraged to purchase a policy but are not required. The rates are lower in these zones as the properties are not directly in the path of inundation from storm surge or storm water. However—and that is a big *however*—residents in moderate and minimal risk zones are affected more and more with each passing year. Recent hurricanes such as Michael and Harvey are revealing cartographical challenges of projected floodwaters. For Harvey in Houston, only 18% of properties that experienced flooding had flood insurance policies. That means that over 80% of structures affected by storm surge and storm water were deemed to be at moderate or minimal risk (though with some exceptions). Maps overlaying flood risk with actual flood damage are astounding in their misalignment—a misalignment brought about by overdevelopment and changing water levels. And this is happening even more frequently. There is urgency behind ensuring that residents in moderate or minimal risk zones purchase policies since flood insurance can be key in helping communities rebuild after catastrophe. Human-built neighborhoods and communities are types of environments in need of protection against erasure, not only because displaced people require new living situations if the homes are not rebuilt but also because leaving neighborhoods behind as derelict might potentially introduce new ecological problems. Low participation in the NFIP is an environmental problem.

To be clear, not all the pressure should be placed on residents and businesses located in moderate- or low-risk areas. Research reveals three general categories of reasons why voluntary participation in the NFIP is so low: distrust, experience, and money. With *distrust* of the maps and science of flood projections, we see that the maps are often wrong, yes, and at times in direct contradiction to the experience so valued by residents (Alfonso et al., 2016). The maps are contentious renderings both in political and scientific senses of the word. Assigning a specific zone to a given neighborhood can have massive impacts on that community's livelihood and ability to thrive (Aerts & Botzen, 2011; Chen, 2018), and often FEMA is not given sufficient resources to update their maps (Scata, 2017), or at least do so in a thorough way that accounts for ever-shifting realities. With every hurricane and tropical storm that wallops the United States' coastlines, we are poignantly reminded of the difficulty in predicting water's flow. Surging waters care little for artificial demarcations; therefore, many might see maps as guesses at best. In terms of *experience*, evidence points to experience as a primary consideration (Osberghaus, 2017; Shao et al., 2017; Tyszka et al., 2017) in assessing risk, as in those who have experienced damage to property prior to an assessment will tend to characterize the future risk as more likely. Those who have no experience, which is most residents, will underestimate the risk. Finally, but not least significantly, is *money*. There are two levels of this. One is large-scale policy, where we need to consider vouchers and other forms of assistance and how premium rates affect specific communities over others and why (this is where the work of Barbara George in this collection comes in as it relates to the problems of "decontextualized" risk from environmental regulatory bodies). One is smaller scale, where premiums just don't work into yearly or monthly budgets for most folks. Research on the latter is complicated, however, as some research has found that cost might not have a lot to do with it and that market penetration rates shouldn't rely on slight premium changes (Dixon et al., 2006).

The dialogic nature of FIRMs and their communicative capacities offer a great deal of potential but also introduce a great deal of uncertainty and resistance. Realistic, grounded work in local communities still needs to be done in addition to larger federal pushes for NFIP participation.

Project Exigence

Coastal Virginia has been lucky so far, with few catastrophic storms walloping its shores. As part of the preparation efforts, the Hampton Roads Planning District Commission (HRPDC) looked at flood insurance policy rates and found real concern. They didn't want us to make the same mistake other regions have. To anticipate these realities for our region, the HRPDC received two related state grants from the Virginia Department of Conservation and Recreation (DCR). The first grant allocated money for HRPDC's Coastal Resilience Committee to design and develop a website called Get Flood Fluent (getfloodfluent.org), an engaging and interactive site aimed at contextualizing the threat of flooding to Hampton Roads residents through a series of questions. The overarching issues explored by the site are *what flood insurance is, why residents need it,* and *how much it costs.* The latter is where the second grant comes in.

The exigence behind the second grant was the lack of clarity on just how much flood insurance costs for those in midlevel risk regions. The underlying assumption made by the HRPDC was that residents of the region did not purchase flood insurance policies because there was a considerable amount of mystery around its cost. Our goal then was to *demystify.* One way to do so was by developing a region-specific calculator that would allow residents to receive an estimate of their yearly premiums without having to contact an insurance provider, their mortgage lender, or FEMA's website. We would be doing that translation (Slack et al., 1993) and centralization work for them. It was a very traditional project in technical communication: take complex insurance rates from FIRMs and the NFIP's rate manual (specifically appendix J) and make the information more understandable for a lay audience. In this sense, our project skirts the complexity of the politics of flood insurance and the psychological factors involved in purchasing it and engages strictly in this question: How does one design a calculator aimed at giving users an estimate on a yearly premium while also breaking down the complex nature of flood insurance and revealing *why* yearly premiums are what they are? (The second part was not part of the grant, but these two lines of inquiry become impossible to disconnect.)

Project Design

The first decision point was what to call the tool. The language used by FEMA was "rate tables," and rate is an accurate term when discussing yearly premium costs and other financials. And even though the grant was focused on the procedural, technical aspect of the calculator itself, I envisioned something more for the tool. The base requirement was figuring costs, but it would be hard to bring users through this process without educating them about scientific and actuarial terms. This is the difference between knowing you are in Zone AE and knowing *why* this designation is different from Zone X. I couldn't justify in my own mind a distinction between these two. So, instead of calling it a flood *rate* calculator, I created a new URL using university server space and titled it officially "Flood *Risk* Calculator." The name stuck.

THE CODING

The second decision point was figuring out how to learn everything about flood insurance and coding and where they would intersect. The conceptual and logical structures of the tool needed to be designed deductively, beginning with the larger divisions and moving "downward" into the more specific features. The key variables were as follows:

- flood zone (e.g., AE, X, VE, etc.)

- location of construction (e.g., city)

- date of construction (e.g., pre-FIRM or post-FIRM)

- residence type (e.g., primary or not)

- occupancy (e.g., single family)

- structure type (e.g., with basement, no basement, mobile home, etc.)

- contents location (e.g., first floor, second floor, etc.)

Other factors included higher risk areas, such as zone VE, or newer homes that have rate tables dictated by the structural level relative to the "base flood elevation" or BFE (e.g., +1 feet, -2 feet). This requires an extra step by users in these regions, taking the BFE of their property and literally adding the number of steps up to the front door.

Turns out, I knew very little about coding (deep down I knew this)—far less than would be required to fulfill the functionality of the tool. And here is why: flood insurance rates are set by a strain of six to eight variables, as listed earlier. They are all logically connected, starting with higher level branching (flood zones) and then moving to lower-level variables (where "contents" are located within your structure). You have to carry the option with you as you select the next one. This meant coding a tool that could output a rate based upon the user's unique sequential responses to the six to eight variables. From my end there were two main categories of options here. One was the long, tedious way that was HTML-based and the other was more elegant and streamlined but JavaScript-based. I spent a great deal of time working on both simultaneously, even though I didn't know JavaScript at all, so I spent hours learning the basic functionality of variable dropdown menu logic. I enrolled in

Codecademy and dwelled in forums.

The core communication related to appendix J in the NFIP manual is helping users locate their rate in the table (see again fig. 4.1). If a user knew exactly all the information about their criteria, they could place their finger on the appropriate box and do the math themselves to figure out how much their premiums would be. The hope here is that this cost would encourage enrollment in the NFIP. And maybe it would if users knew how to get this calculation. But what if the homeowner doesn't know what any of these variables mean? Doesn't know what a "non-elevated with subgrade crawlspace" is, or if they have one? There's an inescapable expository component to this calculation, and it was leading me toward more HTML-based coding because I thought I needed more screen space for textual information. If I were to have this be a sequence-based tool, then each "step" would have its very own page, giving me ample space to explain flood zones and building types. I also needed to embed the searchable GIS tool provided by FEMA's Mapping Service Center (MSC), which is the main tool used to learn flood zone designation. Ultimately, when I presented my HTML-based prototype (fig. 4.3) to the HRPDC team, they thought it clunky and requested more streamlining. When I presented my JavaScript version (fig. 4.4), it was clear that it lacked a positive aesthetic—and I agreed. I didn't have time to work on the cascading style sheet (CSS) since I had to learn JavaScript and write the expository text.

The time spent on the technical aspects was worthwhile. The process of coding was the lens through which I learned about flood insurance policy and its actuarial framework. The coding forced me to focus on the relationships of each variable and how the FIRMs were only the beginning of how flood insurance is communicated. I could not have learned how to communicate about flood insurance policy without first trying to do the coding work because I knew how deeply logical the system was and how dependent each variable was on the others. I learned quickly why flood insurance is so intimidating to residents; I began to better understand the problems inherent in increasing flood insurance policies for those who are not required to but should have them, and the complexity involved in framing the problem as both a protection of personal property and a communal good.

This explains why the flood risk calculator lives on the Get Flood Fluent website, since the cost and the understanding are so intertwined. To integrate the tool I envisioned into the website—and because building it was a bit beyond my ability—the HRPDC involved a designer and coder

Figure 4.3. Screenshot of HTML-based prototype of the flood risk calculator.

Figure 4.4. Screenshot of JavaScript-based prototype of the flood risk calculator.

used to develop the larger website. The developer introduced me to shiny things like PHP code and AJAX programming, which allow for the type of "memory" I was looking for in terms of bringing variables through. I worked in conjunction with this contracted coder and was able to spend my final months on the website's actual written components.

THE COMMUNICATION

While my nightmares about JavaScript declined, I still had a great deal of work to do in terms of composing web text that up to this point had been dappled and scrappy. But it became easier knowing now that the calculator would be within a nice-looking web ecosystem and transition from step to step smoothly and without having to load a page. Here are each of the steps and the communicative challenges within them:

Landing page. This page included a disclaimer, a description of the tool, why it is called a "risk" calculator and not "rate" calculator, and for whom it was designed. We used the following language: "We call this tool a 'Risk Calculator' because in addition to providing an estimate on flood insurance for your property and belongings, the tool also provides insight along the way as to what specific terminology means and why differences in NFIP rate table risk categories matter."

Flood zone. This one was tricky. This is the page where the GIS property tool needed to be embedded. We needed, therefore, to instruct the user on how to locate their address, select their property footprint, and then find the flood zone designation (three separate steps on the same tool), but then also describe what flood zones are and what the letters mean. Explaining the difference between 100- and 500-year floods was also a critical part of this because should the user go back to the FEMA MSC, the map layers use those percentages (1% and 0.2%, respectively). For some significant context, we also thought it useful to share the following about 500-year floods after describing what the term means: "For example, Houston alone has experienced multiple 500-year flood events just in the past few years."

Location and date. These two steps are separate but connected. Knowing where the property is located is significant because we would need to know if there is eligibility for a discounted rate for those living in cities participating in the Community Rating System (CRS). It also forks into the next step where the date the structure was built will be included. When the property structure was built matters a great deal because while the NFIP's FIRMs rolled out in 1974, not every locality enrolled in the program right away. So, homes in the same town built in 1954 and 1988 will have different rates. We decided not to go into too much depth on why 1974 was so key (properties built after that, "post-FIRM," required more mitigation considerations if built in NFIP-participating localities). We identified this as secondary in terms of what the user needed to know at this time.

Primary versus nonprimary. The "what" of this step was straightforward (you either live in a residence for half a year or you do not). The "why" part indicated that, typically, nonprimary homes are vacation homes that are not always occupied and/or are situated in riskier environments. We wanted to communicate these points without making it seem that all nonprimary homes were just fancy vacation homes.

Occupancy type. Things get a bit more complicated here as the user needs to do a bit more thinking about the nature of the property. We used icons deliberately from here, starting with using them to indicate differences between single family and two- to four-family homes, as well as residential other (e.g., condo) and nonresidential business. In the description, we were also sure to include how much of the property can be used for business.

Structure type. This step is almost entirely visual, as the only text we include is that homes with basements have higher rates. The images (fig. 4.5) were designed to help the user approximate the type of structure they are trying to insure, the existing text out there describing the differences between, say, "elevated on crawlspace" and "elevated with subgrade crawlspace" was just not very helpful.

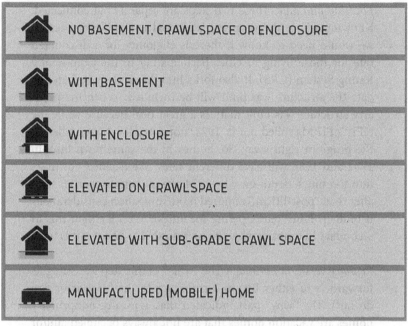

Figure 4.5. Icons describing structure type within the risk calculator.

Elevation. This was by far the most challenging topic. Luckily, it only applies to certain zones (like AE) and only to structures built after 1974. In appendix J, this table is configured entirely different, with feet above or below the base flood elevation (BFE) being the main variable. I won't go into too much detail about how the math behind this works, but this step requires the user to go back to the parcel map and find what their "Ground Height Relative to BFE" number is, and input it—and then, input the number of steps up to the first floor of their property. In this step, we have the technical how-to language, but we also had to describe BFE. Here is the text:

Some properties in higher risk zones built after Flood Insurance Rate Maps were drawn require one more piece of information: "Elevation of Lowest Floor Above or Below the BFE." Yours is one of those properties.

BFE stands for Base Flood Elevation, which FEMA defines as the computed elevation to which the flood is anticipated to rise during the base flood. The base flood is also referred to as the 1-percent annual chance flood or 100-year flood.

We need to know the relationship between your BFE number and the height of your first floor. Our formula will do the math for you—you just need to tell us those two numbers! So, first use the map again. Click on your footprint again like you did in step one. Press the arrows until you see the number for Ground Height Relative to the BFE. Type that number into box 1.

Next, tell us how many steps up from the ground your first floor is. Put that number in box 2. (Note: The formula will automatically multiply your number by 0.625 feet, which is around the average height of a step.)

This language is vital because it is the most complex step and the most likely step where we would lose users. Having to go back to the GIS tool, input a decimal number (1.24 or some such), and then the number of steps up to their first floor is quite a bit to ask. The alternative is getting a BFE certificate, but those can be costly. If we haven't lost users yet with this task, they can move to the final step.

Coverage Calculations. Unless users are in a V or VE zone or need to use the BFE calculator, they are done and can now choose the amount of coverage they'd like for both structure and contents. The coding here is elegant enough not to "split" these pages into two, which I thought would need to be done at first. Users can select their coverage options and receive the output for a yearly premium. The explanatory text here covers why structure and contents are separate, what users can do if they cannot afford it, and what the next steps are. We hoped this text would move users from the estimate part to contacting an insurance agent. Making this connection is a critical component to getting to our intended solution, as an estimate without an action is inadequate for the larger goals of this project. All the language in and previous to this section is foundational to solving our problem since any confusing text can lead users to feel less confident inquiring about flood insurance. If users don't contact an insurance agent, then our tool did not do its job.

The coding and communication for the tool is now complete, and I invite readers to visit the website and tool and provide feedback (getfloodfluent. org). It is really pretty—prettier than I could ever have imagined (fig. 4.6). And the language, I believe, is clear and helpful. But more testing will need to be done.

In all, a great deal of work remains to be done to ensure residents are continually protected against the existential threat of climate change and the hurricanes and tropical storms that are increasing in strength and quantity. Much work needs to be done to ensure that the risk of flooding does not disproportionally affect one socioeconomic or racial group over another. And work is required to develop and maintain the actuarial framework of the NFIP (which has changed as of October 2021) as well as the modeling of future floods. I am hopeful that the climate scientists, public policy and zoning experts, insurance adjusters, government employees, cartographers, and modeling and simulation experts engaging in this

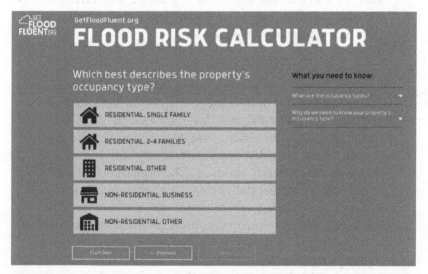

Figure 4.6. Screenshot of typical desktop view of the final design of the flood risk calculator.

great deal of work are doing the best they can to solve these problems. Most of this work is beyond the disciplinary scope of what I have been trained to do and quite frankly scientific and mathematical in ways that limit my participation in them at higher levels.

That said, there is power in doing things such as translating tables from appendices in documents largely unread by the public. The directionality is just different. It might feel disempowering to take work done by those who get paid more than you and try to filter it through a nonexpert's lens. But from my end the logic was different. If I could get more residents of my community to purchase flood insurance policies, then their homes would be protected—that is obvious enough. But more policies purchased also means lower premiums for all residents, which feeds itself into more policies purchased. And the higher the rates of policy purchasing and market penetration in my community, the more likely we are to maintain our neighborhoods and all the things that connect us. And the more we're able to maintain our neighborhoods, the more we'll keep optimism high. Without flood insurance policies, people will lose their homes regardless of the recent article in climate science or adjustments to projected temperatures. Our communities are environments and protecting them and preparing them for a rebuild is a form of environmental action. Demystifying flood insurance and its costs is one small solution we in technical communication can use our expertise in plain language and user-centered design to address.

Conclusion: A Field-Based Assessment of Flood Insurance

You'll notice that there wasn't really an in-depth technical communication framework established early in this chapter. The truth is that the grant project moved quickly, and a good deal of technical work needed to be done. I didn't get to frame the project and its language. I was tasked with creating a calculator within an assigned grant project and had 18 months or so to do it. I didn't consult any technical communication theory, didn't weigh different scholarly approaches. I had to learn as much as I could about flood insurance and coding. I didn't have time for anything else. I just had to build *the thing*. I was not fulfilling an "appropriate role" (Blythe et al., 2008) for researchers in technical communication, namely one that supports the current strategies in place by its residents in the midst of wicked environmental problems. But then again, was this project an example of participatory action research? I don't think so, but the goal was still bringing about positive change in the resilience of the community. And maybe it was an "appropriate role" because I was enacting the current strategies of a regional governmental authority. Is that the same or different from engaging in public, resident-based strategies? I'm not sure, honestly.

In hindsight I can see how my technical communication compass always pointed the calculator toward education. Using the calculator as a platform for plain language was not part of the grant, but it is what happened. I found it difficult just to create a calculator without bringing the user through each step, ensuring they had enough clear information to select from their menu options. The team had to eventually bring in a design savior to help me with the aesthetic of it, but I do think that spending time on language was important—more important than being able to claim that I did this all by myself. My role as a technical communicator in this project, then, was really one of just filling in some gaps. Nothing too extravagant, but a movement of the compass toward education and plain language, nonetheless. Even in the face of overwhelming global challenges such as flooding, great potential remains for rhetorical and even measurable change within rather banal and unassuming technocratic spaces.

I'm left pondering: What is a technical communicator's role in this cartographical, actuarial, and political ecology? Is it in the scientific deliberations in modeling? The perceptual changes in psychology? The communication of maps? After the project, I am even more convinced that flood insurance could and should be a site of scholarship for our field. I'm inspired by the work of Cagle and Tillery (2015) and their extensive interdisciplinary research into climate change communication, synthesizing an expanse of research to help TC scholars better integrate outside scholarship into our work. On the topic of flood insurance, I propose that we might invert this exigence and think about how we in TC can bring our own scholarship to others. I'd like to briefly lay some groundwork to get us going based upon my claims in the opening paragraph of this chapter.

CARTOGRAPHICAL RHETORICS

Monmonier's work on the NFIP and the flood insurance rate maps (FIRMs) is second to none. But his book was written over 20 years ago, and before FEMA digitized their flood rate maps. An updated understanding of the digital components of the searchable Mapping Service Center (MSC) might be justified, especially since one of the core findings of the National Research Council (NRC)'s publication *Mapping the Zone: Improving Flood Map Accuracy* (2009) was: "Finding 5: FEMA's transition to digital flood mapping during the Map Modernization Program creates opportunities for significant improvements in the communication of flood hazards and flood risks through maps and web-based products" (p. 5). Even amid the in-depth analysis of base flood elevations and topographical data layers required to improve modeling, simulation, and zoning in the NFIP, communication about flood risk remains a core need, and one acknowledged by the scientific and cartographical communities. Testing and theorizing the digital maps and their augmentation-based exposition of concepts and mapping terms presented through FIRMs would aid the field of cartography in their communicative efforts, which continually need UX testing (see Kostelnick et al., 2013). Our expertise in thinking through power relationships, representations, ideology, and digital rhetorics would apply well here and fill a gap. It might also be interesting to think about how a resident can argue for their property to be redesignated in a letter of map amendment (LOMA), literally a change to the map.

Environmental Plain Language

The ideas of 100-year floods, 500-year floods, and base flood elevation (BFE) were quite complicated to think about how to communicate. Even the mayor of Houston in the aftermath of Harvey tried his hand at integrating these concepts into public discourse, with varying degrees of success. There is much work to be done in applying basic plain language principles to flood insurance, perhaps using Russell Willerton's (2015) book or the work of Jones et al. (2012) as guides. The work of Miriam F. Williams (2010) and Richard Rothstein (2017) might also be instructive here in thinking through the "veiled" ways that maps serve as versions of regulatory writing themselves, of realities that are inherently political and in service to mostly White communities. Merging mapping discourse with deconstructed visions of regulatory writing might prove useful, especially within scholarship in economic psychology that studies the value of communicative framing for increasing participation and those willing to pay for flood insurance (Botzen et al., 2013). Is plain language even the best communicative frame to connect with residents' existing schemas?

ZONING REGULATIONS AND ENVIRONMENTAL JUSTICE

Popular media often covers how changes to flood insurance zones, especially in New York City (Chen, 2018), will play a significant role in the future livelihood and affordability in affected neighborhoods—in the case of New York City, this might be the Jamaica Bay area. Within such coverage are conversations pointing to the lack of programs for helping poor communities pay for policies (Fields, 2020; Kailath, 2016). Taken together, these threads can be woven into Nixon's (2011) environmental justice work on "slow violence," a concept to describe the often incremental and invisible environmental catastrophes that take place over time and that often disproportionly affect lower-income populations. Conversely, these threads can also be woven into a tapestry of political science, specifically work that investigates the fundamental responsibilities of residents and of the government (Elliott, 2017, 2019). Either way, the role of politics in mapping and zoning connects directly to TC's knack for cracking the objectivist sheen of regulatory writing. Projects highlighting such political realities are in great need.

RISK LITERACY AND COMMUNITY ENGAGEMENT

It has become a truism that effective risk communication strategies that lead to desired action are localized, with messages tailored to the community being addressed. This affirms what we in the field think to be theoretically sound, rhetorical approaches to risk (Grabill & Simmons, 1998), but what we *might* extend our work to do are audience analyses of specific regions on flood risk and the role of insurance. Flooding in the Midwest is different from flooding in coastal Virginia, and even within communities in Virginia, messaging can be different. Other research that can be locally situated is behavior analysis research, thinking about how social networks provide effective avenues for risk communication (Haer et al., 2016), specifically how social networks can alter risk perceptions (Petrolia et al., 2013). Our field's ability to understand the complexity of communications within networked environments would contribute well to these areas of inquiry.

The stakes are high and urgent. And to be sure, the reasons why residents opt not to purchase a flood policy are numerous and enmeshed in a complex web of psychological and socioeconomic factors. But these factors do not—and *should* not—minimize the work of technical communicators working in governmental and nonprofit contexts who initiate dialogue with regional residents that simplifies what flood insurance is, how much it costs, and why it costs what it does. Flood insurance is one of the primary tools available for communities to increase their resilience and prepare for recovery after the inevitable; the work described here is one solution to the environmental problems surrounding community resilience in rebuilding post-disaster. I have used this chapter to argue that these factors are precisely *why* initiating dialogue is imperative, indeed urgent, and that the work presented here is indeed a form of environmental action falling under the purview of technical communication. This chapter has hopefully made plain the enlivening opportunities around the persistently mundane in technical communication while also opening possibilities for more research in the exhilarating world of flood insurance.

References

Aerts, J. C., & Botzen, W. J. (2011). Flood-resilient waterfront development in New York City: Bridging flood insurance, building codes, and flood zoning. *Annals of the New York Academy of Sciences, 1227*, 1–82. https://doi.org/10.1111/j.1749-6632.2011.06074.x

Alfonso, L., Mukolwe, M. M., & Di Baldassarre, G. (2016). Probabilistic flood maps to support decision-making: Mapping the value of information. *Water Resources Research, 52*(2), 1026–1043. https://doi.org/10.1002/2015WR017378

Barton, B. F., & Barton, M. S. (1993/2004). Ideology and the map: Toward a postmodern visual design practice. In J. Johnson-Eilola & S. A. Selber (Eds.), *Central works in technical communication* (pp. 232–252). Oxford University Press.

Blythe, S., Grabill, J. T., & Riley, K. (2008). Action research and wicked environmental problems: Exploring appropriate roles for researchers in professional communication. *Journal of Business and Technical Communication, 22*(3), 272–298.

Botzen, W. J. Wouter, de Boer, J., & Terpstra, T. (2013). Framing of risk and preferences for annual and multi-year flood insurance. *Journal of Economic Psychology, 39*(December), 357–375. https://doi.org/10.1016/j.joep.2013.05.007

Buley, L. (2013). *The user experience team of one: A research and design survival guide*. Rosenfeld Media.

Cagle, L. E., & Tillery, D. (2015). Climate change research across disciplines: The value and uses of multidisciplinary research reviews for technical communication. *Technical Communication Quarterly, 24*(2), 147–163. https://doi.org/10.1080/10572252.2015. 1001296

Chen, D. W. (2018, January 7). In New York, drawing flood maps is a 'Game of Inches.' *New York Times*. https://www.nytimes.com/2018/01/07/nyregion/new-york-city-flood-maps-fema.html

Dedman, B. (2014, February 18). Why taxpayers will bail out the rich when the next storm hits US. *NBC News*. https://www.nbcnews.com/news/investigations/why-taxpayers-will-bail-out-rich-when-next-storm-hits-n25901

Dixon, L., Noreen, C., Seabury, S. A., & Overton, A. (2006). *The National Flood Insurance Program's market penetration rate: Estimates and policy implications.* Rand. https://www.rand.org/pubs/technical_reports/TR300.html.

Elliott, R. (2017). Who pays for the next wave? The American welfare state and responsibility for flood risk. *Politics and Society, 45*(3), 415–440. https://doi.org/10.1177/00323292 17714785

Elliott, R. (2019). "Scarier than another storm": Values at risk in the mapping and insuring of US floodplains. *British Journal of Sociology, 70*(3), 1067–1090. https://doi.org/10.1111/1468-4446.12381

Fields, S. (2020, February 4). "Where it can rain, it can flood." Still, most Americans do not have flood insurance. *Marketplace*. https://www.marketplace.org/2020/02/24/where-it-can-rain-it-can-flood-still-most-americans-do-not-have-flood-insurance/

Grabill, J. T., & Simmons, W. M. (1998). Toward a critical rhetoric of risk communication: Producing citizens and the role of technical communicators. *Technical Communication Quarterly, 7*(4), 415–441.

Haer, T., Botzen, W. J. W., & Aerts, J. C. J. H. (2016). The effectiveness of flood risk communication strategies and the influence of social networks: Insights from an agent-based model. *Environmental Science and Policy, 60*, 44–52. https://doi.org/10.1016/ j.envsci.2016.03.006

Ingraham, C. (2017, August 29). Houston is experiencing its third "500-year" flood in 3 years. How is that possible? *Washington Post*. https://www.washingtonpost.com/news/wonk/wp/ 2017/08/29/ houston-is-experiencing-its-third-500-year-flood-in-3-years-how-is-that-possible/

Jones, N., McDavid, J., Derthick, K., Dowell, R., & Spyridakis, J. (2012). Plain language in environmental policy documents: An assessment of reader comprehension and perceptions. *Journal of Technical Writing and Communication, 42*(4), 331–371. https://doi.org/10.2190/TW.42.4.b

Kailath, R. (2016, September 30). New maps label much of New Orleans out of flood hazard area. *National Public Radio (NPR)*. https://www.npr.org/2016/09/30/495794999/new-maps-label-much-of-new-orleans-out-of-flood-hazard-area

Kousky, C. (2016, April 6). *Flood insurance: Why don't people buy it?* Public Policy Institute of California. https://www.ppic.org/blog/flood-insurance-why-dont-people-buy-it/

Kostelnick, J. C., McDermott, D., Rowley, R. J., & Bunnyfield, N. (2013). A cartographical framework for visualizing risk. *Cartographica, 48*(3), 200–224. http://dx.doi.org/ 10.3138/carto.48.3.1531

Michel-Kerjan, E., Botzen W., Kunreuther, H., Atreya, A., et al. (2015). *Why many individuals still lack flood protection: new findings*. Zurich Insurance Company and Wharton School of the University of Pennsylvania. https://www.zurich.com/en/knowledge/topics/flood-resilience/why-people-dont-buy-flood-insurance

Monmonier, M. (1997). *Cartographies of danger: Mapping hazards in America.* University of Chicago Press.

National Research Council (NRC). (2009). *Mapping the zone: Improving flood map accuracy.* National Academies Press. https://doi.org/10.17226/12573.

Nixon, R. (2011). *Slow violence and the environmentalism of the poor.* Harvard University Press.

Osberghaus, D. (2017). The effect of flood experience on household mitigation: Evidence from longitudinal and insurance data. *Global Environmental Change, 43*, 126–136. https://doi.org/10.1016/j.gloenvcha.2017.02.003.

Petrolia, D. R., Landry, C. E., & Coble, K. H. (2013). Risk preferences, risk perceptions, and flood insurance. *Land Economics, 89*(2), 227–245.

Pralle, S. (2017, September 7). Hurricane Harvey shows how floods don't pay attention to flood zone maps—or politicians. *Washington Post.* https://www.washingtonpost.com /news/monkey-cage/wp/2017/09/07/hurricane-harvey-shows-howfloods-dont-pay-attention-to-flood-zone-maps-or-politicians

Propen, A. (2007). Visual communication and the map: How maps as visual objects convey meaning in specific contexts. *Technical Communication Quarterly, 16*(2), 233–254.

Rothstein, R. (2017). *The color of law: A forgotten history of our how government segregated America.* W. W. Norton.

Scata, J. (2017, October 12). *FEMA's outdated and backward-looking flood maps.* Natural Resources Defense Council (NRDC). https://www.nrdc.org/experts/joel-scata/femas-outdated-and-backward-looking-flood-maps

Shao, W., Xian, S., Lin, N., Kunreuther, H., Jackson, N., & Goidel, K. (2017). Understanding the effects of past flood events and perceived and estimated flood risks on individuals' voluntary flood insurance purchase behavior. *Water Research, 108*, 391–400. https://doi.org/10.1016/j.watres.2016.11.021

Slack, J. D., Miller, D. J., & Doak, J. (1993). The technical communicator as author: Meaning, power, authority. *Journal of Business and Technical Communication, 7*(1), 12–36. https://doi.org/10.1177/1050651993007001002

Tyszka, T., Zielonka, P., & Rzadca, R. (Eds.). (2017). *Large risks with low probabilities: Perceptions and willingness to take preventive measures against flooding.* IWA.

Welhausen, C. A. (2015). Power and authority in disease maps: Visualizing medical cartography through yellow fever mapping. *Journal of Business and Technical Communication, 29*(3), 257–283.

Willerton, R. (2015). *Plain language and ethical action: A dialogic approach to technical content in the twenty-first century*. ATTW series. Routledge.

Williams, M. F. (2010). *From black codes to recodification: Removing the veil from regulatory writing*. Baywood's Technical Communication book series. Routledge.

Chapter 5

Collaborating for Clean Air

Virtue Ethics and the Cultivation of Transformational Service-Learning Partnerships

LAUREN E. CAGLE AND ROBERTA BURNES

Introduction

In 2017, a chance meeting between this chapter's authors—an environmental educator and a rhetoric professor—at a March for Science led to a multiyear service-learning collaboration in which the Kentucky Division for Air Quality (DAQ) has functioned as a partner for a series of advanced technical communication (TC) courses at the University of Kentucky. In these courses, undergraduate students learned about air quality—a technical, environmental topic—and then translated that knowledge for specific audiences and purposes using a variety of media. Students gained real-world experience creating technical communications for a client, and in turn, the Division for Air Quality has benefited from student-produced brochures, story maps, displays, and videos aimed at the public, as well as internal documents that help government employees communicate more effectively and inclusively for environmental action.

The core story of this collaboration is not about successful student production of deliverables, however, even though those are the most visible concrete outcomes. This collaboration has yielded fruit well outside of the classroom, for us authors, for our students, and for the various organizations that are stakeholders in our work. Ours is a story of relationship building, and of the coproduction of knowledge made possible by that process. In this way, this story is somewhat unusual in the technical communication literature on service learning, which more often centers on student agency, motivation, and outcomes and, to a lesser extent, community impacts and curricular design (e.g., Barton & Evans, 2003; Batova, 2020; Bay, 2017; Bourelle, 2014; Brizee et al., 2020; Jones, 2017a; Matthews & Zimmerman, 1999; Walsh, 2010). Although our story is unusual, or perhaps because it's unusual, we need to tell it. Environmental issues are widely and frequently understood as wicked problems (Stahl, 2014), whose complexity and ambiguity require equally complex and potentially roundabout approaches to analyzing and solving them. As such, valuing collaboration and cocreation of knowledge must become a habit of mind for those interested in working to overcome environmental degradation by acting; environmental action can be simple, but it can rarely be broadly impactful when limited to a single person, organization, or discipline.

We thus offer here an autoethnographic reflection on our partnership, situating our narrative in conversations both in and out of technical communication about the purposes and processes of service learning and community engagement. Narrative more broadly is an established methodology in technical communication (Jones, 2017b). Autoethnography is a specific form of narrative inquiry developed in response to critiques of objectivist research that falsely purports to eliminate the personal and associated biases from scholarly inquiry (Chang, 2016; Choi, 2016; Ellis et al., 2021). As the portmanteau suggests, autoethnography "combines characteristics of autobiography and ethnography," with authors "retroactively and selectively writ[ing] about past experiences" to concomitantly "study a culture's relational practices, common values and beliefs, and shared experiences for the purpose of helping insiders (cultural members) and outsiders (cultural strangers) better understand the culture" (Ellis et al., 2021, pp. 275–276). In this chapter, we do precisely that by retroactively piecing together and reflecting on our shared experiences and connecting those reflections to the practices, values, and beliefs about service learning held within the cultural milieu of academic technical communication. The chapter's structure is a little unorthodox: in addition to the usual scholarly moves of analysis, citation, and theory building, authored primarily by Cagle, we also include snippets of narrative by Roberta, folding in her perspective to create a richer polyvocality.

By happy accident, through our partnership we have stumbled onto many of the best practices and useful critical orientations recommended by scholars both in technical communication (TC) and in the interdisciplinary space of community engagement scholarship, which investigates both engaged teaching and research. In reality, though, what our autoethnography reveals is that it *wasn't* really an accident. Cagle's training in teaching context-specific and genre-based technical communication curricula, as well as her background in feminist theory, critical disability studies, and engaged research methods such as participatory action research, helped guide our path. Her community-engaged research, with its emphasis on asset-based inquiry as an approach to coproduction of knowledge, has also been influential. Perhaps most importantly, though, our shared commitments to certain virtues, above and beyond this partnership, enabled us to build a transformational relationship (Clayton et al., 2010). These virtues include humility, patience, respect, honesty, and curiosity, all of which, as it turns out, are professional assets for both of us in our roles as educators concerned with environmental action, albeit action taken in quite different contexts (i.e., public higher education and state government).

ROBERTA: As a lifelong environmental educator, I have a strong curiosity about the natural world and a desire to spark that curiosity in others. At the Kentucky DAQ, I am responsible for conducting outreach to schools, administering our website, and producing a variety of media to help Kentuckians understand their role, and our agency's role, in protecting air quality. Doing what I do within the confines of a government agency requires patience, respect, and humility, not only with those who seek assistance with air quality concerns, but also with the limitations and constraints inherent in state government. How can I empower people to engage in environmental decision making? How can I guide them to find the data and information they seek? These questions drive me as I tackle the daily challenge of communicating the division's work.

The combination of our prior training, experiences, ethical commitments, and personal habits led us to build a service-learning partnership that is resilient, transformational, and concerned with value beyond just producing student service hours. Our autoethnographic narrative walks through these influences, concomitantly describing our partnership's origins and arguing for the value of such transformational, critically informed collaborations in service of environmental action. What we discovered through the autoethnographic process is that our relationship—how we've built it, what role it plays in our pedagogical work, and what elevates it to the status of a reciprocal partnership that enables coproduction of knowledge—owes as much to the traditions of participatory action research and nonformal environmental education as to service learning. In fact, through the reflective process of writing this chapter, Cagle has shifted to describing our collaboration as "participatory action *teaching*," which encompasses and extends the boundaries of service learning. It is a messier, more expansive thing than service learning, allowing us to keep the emphasis on the teaching, but being very clear that the teaching does not happen without the participatory action outside and sometimes lightyears away from the classroom.

In the remainder of the chapter, we begin with the story of our collaboration, then turn to two key questions that animate the ethnography part of our autoethnography: 1) What are we doing, and 2) how are we doing it? To answer those questions, we work through three key conceptual frameworks: coproduction of knowledge, transformational relationships, and virtue ethics. We end with a reflection on what all this talk of service learning in technical communication means—or might mean—for environmental action. To an extent, we conclude, evaluating action's value is a difficult task that's not always best served by obvious metrics, like tons of waste diverted or percentage increase in students' environmental knowledge. But it is a task that nonetheless demands to be done, so we end with an effort at stating what environmental action means and does in the autoethnographic narrative we present here.

The Story of Our Collaboration

In early 2017, a grassroots national movement by scientists and science supporters resulted in the March for Science. That first year, the March for Science consisted of public protests on Saturday, April 22, in more than 600 cities worldwide (Fleur, 2017). Among those cities was Lexington, Kentucky, where hundreds of people turned out in support of science despite an unseasonably cold and rainy day (Kocher, 2017). Cagle, in her first year as a faculty member at the state's flagship school, the University of Kentucky, was a featured speaker at the march on the pressing issue of climate change and how to overcome denial of it. Roberta, in attendance at the march, was struck by Cagle's speech, which aligned with her own personal and professional interests in climate and environmental advocacy. Despite the size of the crowd milling around as the march concluded, Roberta was able to find Cagle and strike up a conversation about our shared interests.

ROBERTA: I was excited to meet Cagle, who had spoken about many of the challenges I'd been experiencing in my own work as an environmental educator, particularly in the realm of climate and air quality communication. I've always seen myself as a science interpreter or translator, so discovering this whole field of work called "technical communication" was such a revelation for me. For one thing, it validated my belief that communicating about air quality is much more challenging than other kinds of environmental education. As we talked, Cagle told me of her upcoming technical communication course, and we immediately began to brainstorm about collaborative possibilities.

That summer, in 2017, we had several meetings, first on the phone and then in person, to lay the groundwork for our relationship and start to figure out how we might collaborate. That fall, Cagle was slated to teach an advanced technical communication course on a topic of her choice and had already decided to focus on the topic of communicating complex information. Given Cagle's choice, our conversations evolved into discussions of what kinds of complex information the DAQ is tasked with communicating and how Cagle's students might contribute to that effort. This early planning for our first collaborative course was thus both relatively product focused and constrained by Cagle's curricular choices. While we worked together to decide which products to have students work on, our collaboration was still limited relative to how it would develop for later courses. As such, the process hewed closer to what Johanna Phelps-Hillen describes as the "predominant charity models of service learning that reinforce the hegemony of the university within the wider community" (2017, p. 114) than to the kind of "feminist community engagement" Phelps-Hillen advocates, and which we ultimately stumbled into.

The following section delves deeper into our collaboration's evolution, so to set that up, we'll provide here a brief description and timeline of the three courses we have codesigned and delivered, with our collaborative relationship and the need for mutual empowerment becoming increasingly central with each course.

Each course has been an advanced technical communication special topics course in the Department of Writing, Rhetoric, and Digital Studies at the University of Kentucky. These courses all fulfill requirements for the Writing, Rhetoric, and Digital Studies major, as well as the minor in Professional and Technical Writing. As a research-focused faculty member in the department, Cagle typically teaches two classes per semester, and has wide-ranging latitude to decide the focus, curricula, and pedagogy of these courses. Our cotaught classes have been small, ranging from seven to fourteen students, largely juniors and seniors. The students have represented a mix of majors, including writing, rhetoric, and digital studies; integrated strategic communications; computer science; English; philosophy; and digital media and design. This context for the courses is important, as their small size and positioning within a Rhetoric Department allow for time-intensive, hands-on, and theoretically advanced teaching practices that may be limited for instructors with a heavier teaching load and who are teaching students less likely to have had other courses in rhetorical theory and the writing process.

Table 5.1 details key information about each of our three service-learning courses, including their curricular foci, the collaborative practices we engaged in, and the deliverables students produced.

Theorizing Our Partnership

As we wrote in the introduction, our collaboration did not start out deeply theorized or informed by the literature we're invoking here. Sometimes engaging technical communication for environmental action means we must *act* first, then consult literature and theorize later. For us, the autoethnographic process of writing this chapter has pushed us to reflect on these connections in relation to existing scholarship. Well, more truthfully, it's pushed Cagle to dig into scholarship, as deep dives into secondary literature are part of her job description in a way that isn't true for Roberta. That's not to say that we haven't both reflected, though, and our various efforts to do things together, both in and out of the classroom, always involve a lot of discussion. To that end, this section aims to combine scholarly theorization of our collaboration with the kinds of naturally occurring conversations that have led us to a point where we have a collaboration worth theorizing in the first place.

Table 5.1. Details about Each of the Three Service-Learning Courses Collaboratively Planned and Taught by Roberta and Cagle

	Communicating Complex Information (Fall 2017)	User Experience and Usability (Fall 2018)	Digital Accessibility for Disabled Users (Fall 2020)
Learning Outcomes	• gain experience working with a client to communicate complex information • develop proficient understanding of locally relevant technical and scientific issue • practice translating complex information for a variety of audiences and purposes	• develop proficient understanding of a locally relevant technical scientific issue • gain experience creating digital story maps • develop basic proficiency in user experience research methods	• develop proficient understanding of critical disability concepts • practice using digital accessibility tools • gain experience producing accessible digital texts and instructing others in digital accessibility
Collaborative Practices	• selecting some course deliverables • selecting DAQ-related readings • occasional class attendance by Roberta • field trips to DAQ locations • Roberta's participation in student-produced video	• selecting all course deliverables • selecting DAQ-related readings • occasional class attendance by Roberta • field trips to DAQ locations • DAQ stakeholders' attendance at students' final presentations	• selecting curricular focus • selecting all course deliverables • establishing course calendar • regular class attendance by Roberta • DAQ stakeholders' attendance at students' final presentations • post-semester workshops presenting Accessibility User Guides to the Kentucky Energy and Environment Cabinet

continued on next page

Table 5.1. Continued.

	Communicating Complex Information (Fall 2017)	User Experience and Usability (Fall 2018)	Digital Accessibility for Disabled Users (Fall 2020)
Student Deliverables			
Public-facing science communications, including:	• table banners • brochures • videos • PowerPoints • surveys	• digital story map • user experience research report • professional presentation for DAQ stakeholders	• accessible versions of existing DAQ PowerPoints, PDFs, and data visualizations • User Guides targeted at Kentucky Energy and Environment Cabinet employees instructing them how to make PowerPoints, PDFs, and data visualizations accessible

We've loosely grouped our reflections under two organizing questions: 1) What are we doing, and 2) how are we doing it? Both questions will be answered by an interstitial story from Roberta and a discussion of the relevant literature from Cagle. Our hope is that you acquire a clear sense of how we've been working together over the last four years in service of environmental action that exceeds the bounds of classroom learning, or even traditional service-learning classroom-client relationships, along with an understanding of how that work ties into various conversations, particularly in service learning and interdisciplinarity scholarship.

What Are We Doing? Transformational Relationships and the Coproduction of Knowledge

This might seem like an odd question this deep into the chapter, but we pose it to ask a fundamental question about what service learning *is*. This question is a natural, perhaps even necessary, precursor to the following question about how we do what we do. I (Cagle) see two primary interpretations for this question. The former is purely descriptive—what is the thing that we are doing—and the latter is normative—why is *this* the thing that we are doing, rather than some other thing. To answer the question, I'll thus make both descriptive and normative moves by describing *what* we have been doing, then explaining *why* we were doing those things at various points in our collaboration.

For higher education professionals, the meaning of the term "service learning" likely seems self-evident. It is learning that provides service to others. Initially, Cagle understood our partnership through the lens of this definition: her expertise and the students in her classes could serve the Division for Air Quality by producing deliverables that the DAQ did not have the time or resources to produce internally. This early sense of our collaboration neatly fit service-learning scholars' Clayton et al.'s definition of a "transactional relationship," which is "instrumental and often designed to complete short-term tasks. People come together based on an exchange, each offering something that the other desires" (2010, p. 7). Transactional is one of the three service-learning relationship types categorized by Clayton et al.; the other two are "exploitative"—indicating a one-way transaction—and "transformational," which refers to "relationships wherein both persons grow and change because of deeper and more sustainable commitments" (2010, p. 7).

Whether the relationships involved are exploitative, transactional, or transformative, service learning is often understood to be synonymous or at least contiguous with community and/or civic engagement; the service to be done is directed at a community and often with civic ideals in mind, such as providing students with experience engaging in democratic action (see Bringle et al., 2012; Hartman, 2013; Jovanovic et al., 2017). Additional terms, from *community service* to *public scholarship*, have been used over the years to get at this cross-disciplinary idea of higher education professionals breaching the walls of the ivory tower to reach out to external communities.

When we first began working together, we weren't concerned about which, if any, of these terms best described what we were doing. Often, we didn't have any need to label the collaboration to discuss and proceed with our transactional relationship. When describing the collaboration to other higher education stakeholders, Cagle often defaulted to calling it "service learning." This term isn't common outside of higher education, however, so we didn't use it for talking with stakeholders in the DAQ, for example. In hindsight, not needing to settle on a label for our work was valuable, because it left us rather unconstrained in imagining what our collaboration might become, beyond a single transactional, service-based semester.

While we imagined our initial teaching relationship in transactional terms, we were concomitantly building a personal relationship of reciprocal engagement. This kind of personal relationship building echoes the sort of community engagement Johanna Phelps-Hillen (2017) calls for in her work on feminist community engagement via service learning. Phelps-Hillen offers a case study of her own collaborative teaching with a community partner, much like this one, and uses it as an example of "engaging *with* communities, rather than *for* communities" (2017, pp. 113–114). Phelps-Hillen's enactment of this shift involved deliberately finding and cultivating a partner interested in coconstructing a course, rather than an organization that "would simply host students for service hours," the form charity-based service learning often takes (2017, p. 121). Unintentionally, we followed in Phelps-Hillen's footsteps, even in that first year of collaboration, by coconstructing course projects for the students in Cagle's advanced technical communication course, so that the organization's needs and the course learning objectives played equal roles in determining what students would do in the course.

While Phelps-Hillen does not describe her experience in terms of Clayton et al.'s typology of relationships, there seems to be strong overlap between the engagement *with* she advocates and their definition of a transformational relationship: "In a transformational relationship, persons come together in more open-ended processes of indefinite but longer-term duration and bring a receptiveness—if not an overt intention—to explore emergent possibilities, revisit and revise their own goals and identities, and develop systems they work within beyond the status quo" (2010, pp. 7–8). As we became more comfortable and secure working together, we transitioned from an early transactional approach to this kind of transformational partnership, in which both Cagle and Roberta's goals and identities are not fixed, but rather open to change based on "emergent possibilities" and increased willingness to move "beyond the status quo."

ROBERTA: Initially I was a bit in awe of Cagle's academic experience, but this shifted as our collaboration progressed. I grew from seeing myself as more of a student, following Cagle's lead, to a confident co-creator with a colleague. Over time, it became clear that there were many groups that could benefit from the insights we were gaining through our collaboration. Conference presentations to government communicators and environmental educators followed, and each one gave us the chance to hone our message for different audiences. Cagle's academic communication style didn't always make sense to me, just as Cagle and students sometimes misunderstood certain air quality phrases that I took for granted. The trust we have built allows us to quickly check in with each other when one of us is unclear in our communication, and it has fueled our creativity as we explore new ways of collaboration.

We'd argue, then, that feminist community engagement is inherently transformational. Moreover, it relies on a particular kind of partnership-based relationship between the instructor and the community member or organization collaborating with them. The exploitative-transactional-transformative taxonomy doesn't capture the distinct characteristics of such a partnership, as a truly collaborative partnership can arguably serve any of these three, but certainly at least both transactional and transformative service learning. In addition to applying that three-part taxonomy to analyzing our collaboration, then, we also adopt Bringle et al.'s (2009) theorization of *partnerships* as a special subset of relationships with three key qualities: closeness, equity, and integrity.

Importantly, a partnership could exist between or among a variety of stakeholders in the service-learning venture. Through their work on service-learning relationships, Bringle et al. (2009) proposed a multis-takeholder framework called SOFAR, which captures five key stakeholder groups and relationships among them: students, organizations in the community, faculty, administrators on the campus, and residents in the community (see fig. 5.1).

The instructor-organization relationship represented in our collabo-ration is, of course, our primary partnership. This is a somewhat unusual focus, as the value of service learning is so often presumed to accrue primarily, if not exclusively, to students. For example, the National Sur-vey of Student Engagement classifies service learning as a "high-impact practice" that has "positive associations with student learning and reten-tion" (NSSE, 2018, p. 2). While student learning and retention are both desirable outcomes, service-learning scholars have rightly pointed out in recent years that service learning could in fact serve a greater number of the stakeholders involved. Moreover, it *should* serve more stakeholders,

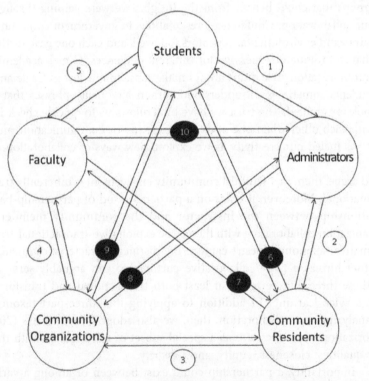

Figure 5.1. SOFAR model of dyadic relationships in service learning.

and ideally in forms outside the limited bounds of charity-based service. This argument is bolstered by Kimme Hea and Wendler Shah's (2016) study of community partners' experiences of service learning, which found that, perhaps counterintuitively, these partners typically don't hold hyperpragmatist views of student products as their primary reason for participating in service learning. Rather, service-learning projects can benefit community partners by providing opportunities for self-reflection, a renewed sense of motivation, or other positive outcomes.

The question of what community partners get from service-learning relationships is an important one, but one that defaults to a transactional response—we know what instructors and students get, and now we also have an expanded view of what community members get in return. This expansion is valuable; it is a move, for example, that allows workplace-influenced TC instructors to shift away from thinking of community partners as clients, in it only for the deliverable. Intentionally applying the framework of transformational partnerships, though, expands the view even further, bringing the relationship between community partner and instructor front and center.

Centering relationships in academic work is hardly a new idea. It is one that animates much pedagogical writing, particularly about instructor-student relationships and peer relationships among students. Cagle is most familiar with a relational focus in scholarship on research methods and management, especially in cultural rhetorics, where feminist and Indigenous development of more ethical orientations to engagement has opened paths for scholars to prioritize relationships among humans and nonhumans alongside or even above data (see Druschke, 2014; Novotny & Gagnon, 2018; Novotny & Opel, 2019; Phelps-Hillen, 2017; Tuck & McKenzie, 2015). These research orientations emphasize co-inquiry and coproduction of knowledge, much like our partnership. The boundaries are thus blurred between transformational partnership-based service learning and research traditions such as participatory action research, engaged rhetoric (Cagle, 2017; Druschke, 2017), and the feminist and Indigenous work cited earlier. Many academics are teacher-researchers, and there's much to be learned and applied from efforts to reimagine how to do research ethically and equitably in partnership with communities. While we did not intentionally build our partnership with these connections in mind, reflecting on what we're doing within the partnership makes these connections clear. In turn, our reflection has allowed us to build a framework for intentional development of habits of mind and relationally oriented virtues that can guide future collaborations, whether they are focused on environmental action or other complex issues that require well-developed partnerships to address. In short, we are arguing that the approach we describe here, while potentially relevant to partnerships focused on any number of issues, is most certainly critical for taking a certain kind of environmental action from within the constraints of the contemporary university.

How Do We Do This? Virtue Ethics as a Foundation for Relationship Building

Stating the value of closeness and long-term transformational relationships in service learning and civic engagement (Clayton et al., 2010; Morton & Bergbauer, 2015) is one thing; actually seeding, growing, and maintaining those relationships is something else altogether. In other words, what does it look like *in practice* to intentionally work toward closeness and transformation in a partnership such as ours? To answer that question, we first argue against the reduction of "practice" to the basic logistics of project and relationship management, then argue instead that our relationship-building practices should be framed by virtue ethics, an ethical framework rooted in the cultivation of personal virtues, practical wisdom, and contextualized decision making.

> ROBERTA: Our relationship has grown from collaboration to partnership largely because of the honesty and respect we both share for each other. Cagle generously gives me the flexibility to attend as many of her class sessions as I want, which affords more opportunities for me and her students to learn from one other. Handing the students materials that I've created and inviting them to deconstruct and analyze those materials has been such a humbling experience! It has allowed me to reflect on many of my practices as an environmental educator and government communicator, not in a self-critical way but in a way that gently encourages me to do better.

The question of "practice" seems deceptively simple. Many resources exist that can guide stakeholders (primarily instructors) on *how* to engage in service learning, and those resources often do address this very question of relationship building and nurturing. That said, to generalize (perhaps a bit unfairly), the practices they focus on are often logistical. For example,

Stanlick and Sell explain that in addition to a practicum course and various institutional reviews, their "partnership process includes frequent meetings to check in and share concerns, lessons learned, new information, and opportunities. Programming is co-planned and supported through grants identified and pursued together" (2016, p. 81). While this article instructs by example, not imperative, other resources are intentionally instructional, such as Bowdon and Scott's extremely practical and detailed book *Service Learning in Technical and Professional Communication* (2002).

Instructing on and modeling logistics is an important part of sharing *how* to do service learning *in practice*. But logistics alone cannot guide us to the kind of relationship we advocate here. The danger of a primarily logistical focus in service-learning instructional literature is that logistical practices need contextualization; they should not be adopted as rigid rules but rather as a heuristic available for context-based application. Take for example the previous description from Stanlick and Sell of their partnership process's "frequent meetings." How frequent? What happens at these meetings? Who is responsible for leading these meetings? How do they ensure that the meetings serve all stakeholders? What processes are in place for them to reflect on and adjust their meeting practices as needed? Answering these questions may be in part a matter of personal preference or external factors; one partner enjoys leading meetings, the other doesn't, or perhaps a work schedule changes, and meeting times must be adjusted to accommodate. But beyond such influences, answering these questions requires stakeholders—in this case, the collaborative partners—to bring shared ways of being, or at least shared ways of *valuing* ways of being, to their collaboration.

To return to our guiding question for this section, then, we argue that *in practice*, intentionally working toward closeness and transformational partnerships requires logistical choices to be rooted in shared habits of mind and behavior that directly serve shared beliefs about and goals for service learning. These shared habits and beliefs are what allow us to apply a virtue ethics framework to our relationship to nurture it into a collaborative partnership. Again, the virtue ethics framework can undergird partnerships focused on a variety of wicked problems, but our specific narrative demonstrates the value of this framework for environmental action in particular.

Such attention to habits of mind and behavior is not absent in the service-learning literature, including the logistically focused literature we cited earlier. For example, d'Arlach and colleagues emphasize the need for reciprocity in service-learning relationships, arguing that university stakeholders must resist the assumption that "the community has a deficit that the resources or expertise of the university can help alleviate," as well as the tendency for service-learning to boil down to students "com[ing] down from the ivory tower to tutor or translate or help the community" (2009, p. 13). Instead, instructors should avoid a charity-based mindset, developing practices motivated and informed by openness, humility, and transparency. In another example, Morton and Bergbauer explicitly tie the pursuit of civic engagement and social justice to "the *practices* of hospitality, compassion, listening, and reflection across social and cultural boundaries" (2015, p. 18). In that list of practices, each is arguably also a habit of mind, or a personal characteristic, with perhaps the exception of "listening."

This list is notably reminiscent of lists of personal virtues, or habits of mind that individuals can cultivate in themselves through everyday practice to develop the wisdom to act virtuously on all occasions. Vallor succinctly describes the "Greek concept of virtue [as] refer[ring] to any stable trait that allows its possessor to excel in fulfilling its distinctive function: for example, a primary virtue of a knife would be the sharpness that enables it to cut well"; when filtered through Plato and Aristotle's examination of ethics, "the concept acquires an explicitly *moral* sense entailing excellence of character" (2016, p. 17, emphasis in original). For Aristotle, this "moral sense" is what makes virtues the basis of "virtue ethics," an ethical framework in which the cultivation of virtues is the primary goal, and the virtues themselves are the primary tool for ethical decision making. The wisdom developed from exercising these virtues in everyday contexts is called *phronesis* (Aristotle, 2012).

Given the apparent connection between the prosocial practices advocated by service-learning scholars and the implementation of virtues through phronesis, we turn here to contemporary scholarship on virtue ethics to reflect on how to intentionally prepare for and cultivate service-learning partnerships like our own. Rather than to-do lists, phronesis gives us solid ground on which to *practice* our ethical decision making (Gallagher, 2018). Virtue ethics gives us a framework through which to evaluate and selectively implement the various logistical practices advocated and described in the service-learning literature.

In their introduction to the 2018 special issue of *Rhetoric Review* on virtue ethics, Duffy et al. argue that contemporary virtue ethics has the flexibility to encompass nontraditional virtues, as "in contexts of oppression and disinformation, for example, skepticism, righteous anger, and resistance are also virtues" (2018, p. 323). So, while the traditional Western virtues belonged to a bounded set, the underlying logic of virtue ethics allows us to add to that set as needed in contemporary contexts. In the case of building reciprocal transformational service-learning partnerships, we advocate cultivating the following five virtues in oneself and the collaboration; phronesis is assumed to be an underlying sixth virtue arising naturally from the exercise of the other virtues. These virtues, we argue, create not only the conditions for an ideal service-learning partnership, but also the conditions for *right* environmental action to emerge from that partnership:

- *Honesty.* The Western virtue ethics tradition provides a particularly useful approach to honesty, especially in contrast to deontological ethical systems, which lead to ethical quagmires in which universal honesty inevitably will result in at least some morally repugnant outcomes. Virtue ethics, with its underlying reliance on the practical wisdom of phronesis and its emphasis on cultivating simultaneous virtues, gives rhetors the opportunity to purposefully deploy honesty toward morally defensible ends. In practice for service-learning partnerships, honesty is a critical virtue motivating stakeholders to be transparent about their capacities and limitations. Moreover, honesty as virtue rather than universal rule allows for tactful decisions about how much to share and when, which can be very helpful, for example, for avoiding information overload during the beginning stages of a relationship. In our relationship, for example, we exercise honesty (and justice) by conversing with each other about our broader beliefs and values, particularly as they relate to ideas about justice and equity, so that we can engage in difficult conversations with students about the challenges of regulating air quality in ways that serve all residents.

- *Generosity.* Generosity is a well-known virtue, made most visible in the practice of giving. What we give isn't necessarily material, of course. We can give grace, understanding, respect, time, the benefit of the doubt, and so on. Bringing the virtue of generosity to bear in a partnership can help create the space necessary to manage the discomforts and challenges inherent to transformational relationships' variability and changes. In our relationship, for example, we exercise honesty and generosity by explicitly discussing communication preferences and habits, including frequency, mode, and purpose of communications, and compromising when our preferences and habits conflict.

- *Respect.* We've written in various places throughout this chapter about resisting charity models of service learning and instead taking an asset-based approach, which is well established in research contexts (see Kramer et al., 2012; McKnight, 2010; as well as Flora et al., 2004, on the community capitals framework). Fundamentally, this shift is about creating the conditions for mutual respect; cultivating the virtue of respect allows one to approach a new relationship with the default assumption that the prospective partner is worthy of respect and that they have assets to bring to the table, whether tangible or intangible. In our relationship, for example, we exercise respect by collaboratively deciding how to introduce Roberta to students and what communication systems will allow them to keep in touch with her but also respect her time.

- *Humility.* Humility is perhaps well understood as the flip side of respect; we should not only look for others' assets but also acknowledge our own deficits. Humility is also often best exercised in conjunction with honesty, as we can protect our partnerships, especially in their fragile early stages, by explicitly sharing with each other our capacities *and* our limitations. The cultivation of humility enables us to practice the important skills of self-reflection and understanding difference, which are key to collaborating across disciplinary or professional boundaries (McGreavy et al., 2014). In our relationship, for example, we exercise humility by telling students when the other member of the partnership is a better resource to answer their questions or guide their work.

- *Justice.* Justice is a complicated virtue (as they all are, though justice perhaps more than most) with long histories of philosophical inquiry in multiple traditions. We adapt Vallor's

(2016, p. 128) conception of justice, which focuses on fair and equitable distribution of benefits and risks, as well as concern for impacts on the basic rights, dignity, or welfare of individuals and groups. Within the context of reciprocal transformational service-learning partnerships, justice plays the important role of guiding the balance of reciprocity and the direction of transformation. As a partnership grows and changes, the partners will necessarily have decisions to make about what projects to take on, what roles to play, and what other stakeholders to involve. Developing the virtue of justice means that, on balance, decisions within the partnership will move toward projects, roles, and stakeholders that fairly distribute benefits and risks in ways that respect others' rights, dignity, and welfare. In our relationship, for example, we exercise justice by voluntarily taking on additional work within the partnership to create and lead workshops for the Kentucky Energy and Environment Cabinet on making digital documents accessible, thereby increasing all Kentuckians' access to regional environmental information.

This virtue ethics framework for a partnership also has the benefit of foregrounding action as a key element. The virtues are not intended to be abstract ideals, but rather guidelines for action; it is only through regular, mundane practice that we can hone our sense of how and when to develop the virtues, as well as how to manage them when they come into conflict. Through that regular action, we cultivate the virtues within ourselves, ideally and eventually cultivating phronesis, or practical wisdom. As an ethical and relational foundation, then, the virtues call us to take action and to take that action thoughtfully. For environmental action, such care can be particularly important, as the complexity of the environment—whatever we understand that to be—means that our actions are necessarily interacting with open-ended systems of which we are just a small part.

Building Participatory Action-Oriented Teaching Partnerships

It is worth noting our focus on the term *action* in the prior section, rather than *activism*, which is perhaps more frequently attached to the modifier *environmental*. While the terms *action* and *activism* are often used interchangeably, there is value to distinguishing them, particularly in this collection, which has only the former term in its title. Making such a distinction is not purely pedantic; rather, it's important to understand what the goals of a reciprocal transformational partnership such as ours are, and those goals might include action or activism or both. Moreover, given our respective positionalities as an instructor at a land-grant university and a professional representative of a government agency, the terms *action* and *activism* do different kinds of work for us.

In distinguishing *action* from *activism*, we borrow from geographers Chatterton and colleagues' (2008) reflections on how participatory action research (PAR) and activism could, but do not necessarily, overlap. Action, they argue, is a central part of PAR, yet many PAR scholars emphasize the "research" aspect of their projects over all else. And research, of course, does require action of some kind—one must *act* to recruit participants, collect data, analyze that data, and share results. However, simply acting in service of research does not necessarily lead to *activist* methodologies or outcomes, which require "a commitment to social transformation, challenging power relations, showing solidarity, recognizing and using emotions, being the change you want to see, and building spaces for critical dialogue" (Chatterton et al., 2008, p. 222). To illustrate especially the social transformation aspect of activism, Chatterton and colleagues cite an example of community engagement around dam construction. Action, they argue, might include things like "using appraisal methods to elicit the views of a community," whereas activism "requires working with that community, to understand, for example, how World Bank policy works, so we can resist and possibly reverse its construction" (2008, p. 218). These elements of activism set a high bar, and frankly, most of the community-based research and teaching done in US higher education does not clear that bar.

This is not a criticism. Rather, it's a caution to not conflate action with activism, thereby misrepresenting the former and devaluing the latter. The service-learning partnership we described in this chapter, however transformational for us personally, and potentially for our students and colleagues, largely does not constitute activism. We are, for the most part, not making deep structural change in response to community-identified ills or oppressive power relations.

But that doesn't mean we aren't acting, or that our action doesn't matter. In fact, we are, and it does. Perhaps we could characterize what we're doing as "participatory action teaching" (PAT), where the emphasis is on the teaching, but the teaching does not happen without the participatory action. This characterization better describes what our partnership has evolved into than does the term *service learning*, which brings along the weight of assumptions about service-as-charity and a historically individualist orientation that fits uneasily with organizational writing. Moreover, PAT is expansive; it's not just about teaching within the confines of a college class, but about a messy ongoing relationship of which our coproduced courses are just one manifestation, and in which there is room for so much more: joint workshops and informal exchanges of expertise and former students becoming organizational collaborators and even the coproduction of knowledge by, for example, our coauthorship of the very book chapter you're reading right now. Moreover, the move from service learning to PAT constitutes a commitment to building a space for transformation. That space begins with individuals who then ripple out into our communities and into the world, and potentially into the environment to drive consequential action.

These actions are consequential. However, we do need to ask whether they're environmentally consequential, given the focus of this edited collection. To answer that question, we first should point out that, while our example is broadly relevant to all kinds of service-learning partnerships, our partnership includes a government agency, which places constraints on what is considered "environmental" and what "actions" are available to us. We must in short, stay on the action side of the action/activism boundary line. This is not a case of taking on a big new environmental issue that no one is yet addressing in an official capacity, as we might see, for example, in environmental justice movements. Rather, our participatory action teaching is one activity joining a constellation of ongoing activities by government officials and other stakeholders aimed at improving and maintaining air quality in the Commonwealth of Kentucky. The air quality here is quite good, and our partnership has not noticeably improved it, certainly not in any quantifiable way.

ROBERTA: It's true, Kentucky's air quality has improved dramatically in the 50 years since the Clean Air Act was passed. Only two areas (Northern Kentucky and Louisville) are classified as not meeting the national air quality standard for ozone pollution. Our work has focused more on improving how we communicate about air quality, and how we make air quality information more accessible to Kentuckians. Some of Cagle's students have been inspired to go further by creating videos and brochures that encourage people to take action (or even change their actions) to protect air quality.

What, then, can we claim is specifically environmental about our actions within our participatory action teaching partnership? There's nothing uniquely environmental about this notion of participatory action teaching. And our partnership is environmental by default because the organizational focus is environmental. But that is enough. Service-learning stakeholders, Stanlick and Sell argue, too often "encourage us to *found* and *create* new programs, initiatives, and research projects rather than to enhance those that already exist" (2017, p. 80, emphasis in original). This encouragement pushes us, instructors and community partners alike, to think of ourselves as superheroes leading the way. But we can, instead, be followers, quietly contributing to ongoing environmental work through the participatory action of teaching. We do not have to be activists, publicly making our stances known, or instrumentalists, measuring every molecule of every air pollutant our work has diverted from the open air, to show that our actions matter. Participatory action teaching has its own value and potential, and that value and potential can serve environmental outcomes, as it has in our case. The real work—the sustainable change—of such projects and the relationships underlying them lies in the transformation of ourselves, moving beyond superficial community engagement to enacting virtue ethics in all our partnerships, whether for environmental action or other complex challenges.

Conclusion

It's perhaps important to note at this point that, despite the thousands of words we just used to describe how beneficial service-learning and related partnerships can be, not every instructor has to want to do this kind of work. Perhaps more importantly, not all instructors have the agency or labor conditions that would allow them to do this kind of work. It takes time and resources, both of which are in short supply for the many higher education instructors in positions of precarity.

Even beyond the logistical issues of time and material resources, participatory action teaching demands extensive attention and care; reflecting on and cultivating the virtues we've espoused here is *work*. And in the case of participatory action teaching, it's work that can be complicated by a thousand things: a mismatch between academic and organizational timelines; the potential for unplanned issues (such as when a student in one of our courses worked for a lobbying firm, so we had to consult counsel about whether she could ethically and legally produce materials for a government agency); and the danger of misaligned expectations in the collaboration, such as when students, instructors, and service-learning partners might have different understandings of and goals for environmental communication.

We cannot provide redress for material inequities with this chapter. But we hope that our autoethnographic example helps anyone—instructor or community partner—interested in finding ways to make service learning part of their repertoire of environmental actions.

ROBERTA: Environmental educators are often the sole providers of public information for their organizations. It's easy to get burned out or settle into comfortable routines when you're the only one doing this work. That's one of the reasons my relationship with Cagle has been so transformative for me. Sure, conferences are great for networking, but it was that first, informal conversation at a community event that opened the door to our partnership and all of its potential. Don't be afraid to step outside your professional bubble and share your elevator speech with people who aren't necessarily in the same line of work. When the opportunity arises, bring your full self to the table.

Doing service learning in the way we've described here will require reflection on what the service-learning partnership aims to accomplish, whether there's room for transformation within the relationship, and what shared virtues will allow the partnership to succeed according to its own terms. But the good news is that that reflection doesn't all have to happen up front; it didn't for us, as you know. Participatory action teaching—the long-term process of finding people with whom you can engage in reciprocal transformation and with whom you can then build pedagogy—is a forgiving process. So, gather your resources, learn from prior action, look at what's already out there, then act.

References

Aristotle. (2012). *Aristotle's Nicomachean Ethics* (R. C. Bartlett & S. D. Collins, Trans.; reprint edition). University of Chicago Press.

Barton, E., & Evans, L. (2003). A case of multiple professionalisms: Service learning and control of communication about organ donation. *Journal of Business and Technical Communication, 17*(4), 413–438. https://doi.org/10.1177/1050651903255303

Batova, T. (2020). An approach for incorporating community-engaged learning in intensive online classes: Sustainability and lean user experience. *Technical Communication Quarterly*. https://doi.org/10.1080/10572252.2020.1860257

Bay, J. (2017). Training technical and professional communication educators for online internship courses. *Technical Communication Quarterly, 26*(3), 329–343. https://doi.org/10.1080/10572252.2017.1339526

Bourelle, T. (2014). Adapting service-learning into the online technical communication classroom: A framework and model. *Technical Communication Quarterly, 23*(4), 247–264. https://doi.org/10.1080/10572252.2014.941782

Bowdon, M., & Scott, B. (2002). *Service learning in technical and professional communication* (first edition). Longman.

Bringle, R. C., Clayton, P., & Price, M. (2012). Partnerships in service learning and civic engagement. *Partnerships: A Journal of Service-Learning and Civic Engagement, 1*(1), Article 1. http://libjournal.uncg.edu/prt/article/view/415

Brizee, A., Pascual-Ferrá, P., & Caranante, G. (2020). High-impact civic engagement: Outcomes of community-based research in technical writing courses. *Journal of Technical Writing and Communication, 50*(3), 224–251. https://doi.org/10.1177/0047281619853266

Cagle, L. E. (2017). Becoming "forces of change": Making a case for engaged rhetoric of science, technology, engineering, and medicine. *Poroi, 12*(2). https://doi.org/10.13008/2151-2957.1260

Chang, H. (2016). *Autoethnography as method*. Routledge.

Chatterton, P., Fuller, D., & Routledge, P. (2008). Relating action to activism: Theoretical and methodological reflections. In S. Kindon, R. Pain, & M. Kesby (Eds.), *Participatory action research approaches and methods: Connecting people, participation and place* (pp. 216–222). Routledge.

Choi, J. (2016). *Creating a multivocal self: Autoethnography as method*. Routledge.

Clayton, P. H., Bringle, R. G., Senor, B., Huq, J., & Morrison, M. (2010). Differentiating and assessing relationships in service-learning and civic engagement: Exploitative, transactional, or transformational. *Michigan Journal of Community Service Learning, 16*(2), 5–21.

d'Arlach, L., Sánchez, B., & Feuer, R. (2009). Voices from the community: A case for reciprocity in service-learning. *Michigan Journal of Community Service Learning, 16*(1), 5–16.

Druschke, C. G. (2014). With whom do we speak? Building transdisciplinary collaborations in rhetoric of science. *Poroi, 10*(1). https://doi.org/10.13008/2151-2957.1175

Druschke, C. G. (2017). The radical insufficiency and wily possibilities of RSTEM. *Poroi, 12*(2). https://doi.org/10.13008/2151-2957.1257

Duffy, J., Gallagher, J., & Holmes, S. (2018). Virtue ethics. *Rhetoric Review, 37*(4), 321–327. https://doi.org/10.1080/07350198.2018.1497882

Ellis, C., Adams, T. E., & Bochner, A. P. (2021). Autoethnography: An overview. *Forum Qualitative Social Science Open Access Repository.* https://nbn-resolving.org/urn:nbn:de:0168-ssoar-363237

Fleur, N. S. (2017, April 22). Scientists, feeling under siege, march against Trump policies. *New York Times.* https://www.nytimes.com/2017/04/22/science/march-for-science.html

Flora, C. B., Flora, J. L., & Fey, S. (2004). *Rural communities: Legacy and change* (second edition). Westview Press.

Gallagher, J. (2018). Enacting virtue ethics. *Rhetoric Review, 37*(4), 379–384. https://doi.org/10.1080/07350198.2018.1497882

Hartman, E. (2013). No values, no democracy: The essential partisanship of a civic engagement movement. *Michigan Journal of Community Service Learning, 19*(2), 58–71.

Jones, N. N. (2017a). Modified immersive situated service learning: A social justice approach to professional communication pedagogy. *Business and Professional Communication Quarterly, 80*(1), 6–28. https://doi.org/10.1177/2329490616680360

Jones, N. N. (2017b). Rhetorical narratives of Black entrepreneurs: The business of race, agency, and cultural empowerment. *Journal of Business and Technical Communication, 31*(3), 319–349. https://doi.org/10.1177/1050651917695540

Jovanovic, S., Moretto, K., & Edwards, K. (2017). Moving from thin to thick conceptions of civic engagement: Faculty identity and goals for democratic learning. *The International Journal of Research on Service-Learning and Community Engagement, 16.*

Kimme Hea, A. C., & Wendler Shah, R. (2016). Silent partners: Developing a critical understanding of community partners in technical communication service-learning pedagogies. *Technical Communication Quarterly, 25*(1), 48–66. https://doi.org/10.1080/10572252.2016.1113727

Kocher, G. (2017, April 22). Hundreds march for science through downtown Lexington. *Lexington Herald Leader.* https://www.kentucky.com/news/local/counties/fayette-county/article146200284.html

Kramer, S., Amos, T., Lazarus, S., & Seedat, M. (2012). The philosophical assumptions, utility and challenges of asset mapping approaches to community engagement. *Journal of Psychology in Africa, 22*(4), 537–544. https://doi.org/10.1080/14330237.2012.10820565

Matthews, C., & Zimmerman, B. B. (1999). Integrating service learning and technical communication: Benefits and challenges. *Technical Communication Quarterly, 8*(4), 383–404. https://doi.org/10.1080/10572259909364676

McGreavy, B., Silka, L., & Lindenfeld, L. (2014). Interdisciplinarity and action-able science: Exploring the generative potential in difference. *Journal of Community Practice, 22*(1–2), 189–209. https://doi.org/10.1080/10705422.2014.901264

McKnight, J. (2010). Asset mapping in communities. In A. Morgan, M. Davies, & E. Ziglio (Eds.), *Health assets in a global context: Theory, methods, action* (pp. 59–76). Springer. https://doi.org/10.1007/978-1-4419-5921-8_4

Morton, K., & Bergbauer, S. (2015). A case for community: Starting with relation-ships and prioritizing community as method in service-learning. *Michigan Journal of Community Service Learning, 22*(1), 18–31.

National Survey of Student Engagement (NSSE). (2018). *NSSE 2018 high-impact practices* (p. 6).

Novotny, M., & Gagnon, J. T. (2018). A shared ownership approach to rhetorical research in trauma communities. *Reflections: A Journal of Community-Engaged Writing and Rhetoric, 18*(1), 71–101.

Novotny, M., & Opel, D. S. (2019). Situating care as feminist rhetorical action in two community-engaged health projects. *Peitho, 22*(1). https://cfshrc.org/article/situating-care-as-feminist-rhetorical-action-in-two-community-engaged-health-projects/

Phelps-Hillen, J. (2017). Inception to implementation: Feminist community engage-ment via service-learning. *Reflections: A Journal of Community-Engaged Writing and Rhetoric, 17*(1), 113–132.

Stahl, C. H. (2014). Out of the Land of Oz: The importance of tackling wicked environmental problems without taming them. *Environment Systems and Decisions, 34*(4), 473–477. https://doi.org/10.1007/s10669-014-9522-5

Stanlick, S., & Sell, M. (2016). Beyond superheroes and sidekicks: Empowerment, efficacy, and education in community partnerships. *Michigan Journal of Community Service Learning, 23*(1), 80–84.

Tuck, E., & McKenzie, M. (2015). *Place in research: Theory, methodology, and methods.* Routledge, Taylor & Francis.

Vallor, S. (2016). *Technology and the virtues: A philosophical guide to a future worth wanting.* Oxford University Press.

Walsh, L. (2010). Constructive interference: Wikis and service learning in the technical communication classroom. *Technical Communication Quarterly, 19*(2), 184–211. https://doi.org/10.1080/10572250903559381

Chapter 6

The Narrative of Silent Stakeholders

Reframing Local Environmental Communications to Include Global Human Impacts

BETH SHIRLEY

Because [the association] is the only hope there is. There is nothing else.

—Participant A

Even as citizens across the United States are beginning to recognize the effects of climate change in their own backyards (Howe et al., 2015), global climate change is disproportionally affecting people in developing nations and has been for some time (IPCC, 2014, 2018, 2021). Yet environmental issues are still often portrayed as requiring a balance between local economic impact against local environmental impact (e.g., Colman & Adragna, 2021). Where does that leave global stakeholders already affected by the impacts of these localized decisions? What is the responsibility of technical communicators to advocate for these disenfranchised communities?

Recent scholarship in technical communication (TC) emphasizes stakeholder engagement, especially for navigating complex local environmental issues and taking environmental action (Cagle & Tillery, 2015; Dodson & Palliser, 2016; Druschke & McGreavy, 2016; Simmons, 2008; Simmons & Grabill, 2007; Walker, 2007). But who counts as a stakeholder in a local policy issue with global ramifications? In practice, stakeholder engagement is almost necessarily limited to including those stakeholders who have access to the technical communicators and policymakers, and whoever has access to participation has power (Simmons, 2008). Because technical communicators can function as mediators in environmental policymaking and communication, we are also in a position to extend access to groups who are traditionally denied it, and we have a responsibility to do so (Hopton, 2013). In the case of climate change and local environmental issues, this must include marginalized groups thousands of miles away who are already experiencing intense effects of anthropogenic climate change (IPCC, 2014, 2018, 2021). How do we in TC determine who counts as a stakeholder, and how do we then make participation in policymaking accessible to those stakeholders who are otherwise inaccessible, invisible, and silent? In this chapter, I argue that the field of stakeholders must include communities in developing nations, even in local environmental policy decisions, and I present the use of societal teleconnections (Moser & Hart, 2015) as a strategy for technical communicators to include these communities in environmental action.

In May of 2017, I took part in a research trip with Dr. Rebecca Walton and Dr. Peg Petrzelka to learn about how women's civil society organizations (CSOs—nonprofit, non-state entities formed locally to improve economic and social opportunities for members) operate in Morocco. One of the most striking memories I have from our research was a member telling us that the greatest benefit to being a member of the association is the women of this village have more agency—more of a voice—when they are part of this collective than they would on their own. Another participant later said one of the goals of their organization is to bring people like us into their village to see what these women's lives are like and to learn about their stories. These were striking comments not only because they drew attention to the women's sense of limited self-efficacy, but also because the comments highlighted the important role TC researchers can have in incorporating the narrative of otherwise silent stakeholders across the breadth of topics and issues that our work encompasses. As the members of the CSO told us more about their goals and challenges, their narratives made one thing clear: these women have a lot at stake in the greater issue of climate change. Yet, because they are marginalized even within their own country for their socioeconomic status, sex, and rural situation, and because they are marginalized on the global scene as citizens of a nation with limited economic interests, they are unheard and not considered in most environmental decisions. Amplifying marginalized voices of these silent stakeholders and connecting their stories to decisions made in the United States can strengthen a technical communicator's argument for environmental action.

As stated above, TC emphasizes stakeholder engagement with science and policy in building more effective rhetorics (Cagle & Tillery, 2015; Dodson & Palliser, 2016; Druschke & McGreavy, 2016; Simmons, 2008; Simmons & Grabill, 2007; Walker, 2007). In practice, we may see that these engagements with stakeholders boil down to a debate between maintaining the current way of life—or a group's livelihood—and protecting the environment. To some extent, this may be a result of the journalistic laws and traditions that give equal weight to opposing sides of an issue (Ceccarelli, 2011). For example, consider the debates regarding protecting coal mining jobs versus protecting the environment by closing coal plants (e.g., Bloomberg, 2017; Hood, 2018). We have known for some time that burning coal and other fossil fuels not only releases harmful pollutants into the immediate air (Querol et al., 1996), but also releases greenhouse gases into the atmosphere contributing to global climate change and creating ripple effects across the planet (Fahey et al., 2017). Amid staunch denial that this is happening, coal mining operations, miner labor unions, and politicians argue that it is more important to protect jobs than to change our energy consumption to mitigate the severity of climate change around the world. The way policies and plans are drafted often reflects this juxtaposition of local and global needs and goals. Of course, replacing coal with solar and wind energies would create more jobs that would be less dangerous and more economically sustainable, so what may be perceived to be at stake locally is not jobs but a way of life, a tradition held by generations of coal miners that formed communities and strong identities within those communities. (Loretta Lynn never sang about being proud to be a solar panel installer's daughter.) Recent resistance to the Green New Deal and other attempts to improve coal communities while cleaning up the environment demonstrate this point (Colman & Adragna, 2021). What is truly being attacked, coal communities may feel, is their culture and their livelihood, their ability to provide for their families and their community in the way they always have.

Current debates in the US about making behavioral changes thus focus on two sides: protecting the local environment against protecting a local way of life and pitting local knowledge and experience against empirical science and data. Consider the argument being made in the photo from the news article in figure 6.1 ("EPA regulations destroy jobs and put seniors at risk"), which implies that protecting the environment means losing jobs. This binary argument is narrow and marginalizing yet paradoxically evolves from stakeholders whose voices are easily heard through collective action (note the United Mine Workers association hats in fig. 6.1).

It is important to recognize the need in the region for economic stability and to understand the local knowledges and experiences of these stakeholders, but technical communicators also need to broaden our parameters for who counts as a stakeholder in these decisions. Who else's

Do environmental regulations reduce employment? Not really.

By David Roberts | @drvox | david@vox.com | Updated Mar 28, 2017, 10:27am EDT

f 🐦 ⤲ SHARE

(Photo by Win McNamee/Getty Images)

On *Fox & Friends* this morning, EPA Administrator Scott Pruitt **made reference** to America's "anti-jobs" environmental regulations.

Figure 6.1. Members of the United Mine Workers of America rally outside of the Environmental Protection Agency headquarters to protest proposed regulations on mining and coal power plants.

economic stability is at stake in this debate about maintaining coal jobs, and what can we learn by listening to other, distant, localized knowledges and experiences?

According to the Intergovernmental Panel on Climate Change (2014, 2018, 2021), developing nations are already being hit hardest by the effects of anthropogenic climate change that are driven primarily by Western actors. Developing nations lack the infrastructure to build resilience and bounce back after an extreme event (such as a hurricane, drought, or desertification, all intensified by anthropogenic climate change). This means that stakeholders in these communities are at elevated risk of the effects of anthropogenic climate change, and because they are not citizens of powerful local communities (local to the US or the Global North), they go unheard and unseen. As technical communicators, we have an imperative to amplify voices of marginalized communities (Hopton, 2012; Walton & Jones, 2013), and narratives can be a strong part of that amplification (Jones et al., 2018). Leaving these silent stakeholders out of the collective narrative and omitting local knowledges and experiences from communications that impact environmental action is a social justice issue.

Critiques of stakeholder engagement cite insufficiencies in how current methods typically conduct that engagement: either that the approach may not open room for negotiation of shared meanings and understandings (Walker, 2007); that processes of participation are still hierarchical and thus value contributions from elite and power-holding groups over other stakeholders (Dodson & Palliser, 2016); or that the deliberation spaces are too complex, either hierarchically or technically (Simmons & Grabill, 2007). Each of these concerns can be complemented by two additional concerns that technical communicators should consider when working in technical communication for environmental action:

1. Existing models of stakeholder engagement appear to situate local and global environmental concerns squarely against local stakeholders' interests (economic or otherwise) when the long-term, global, environmental impacts are going to affect local stakeholders just as much as they impact those who are far away.

2. Existing models of stakeholder engagement fail to consider the narrative of silent stakeholders impacted by decisions made in the US without any input or consideration from those without a voice.

The case study I present here shares the narrative of some of these stake-holders to demonstrate how such narratives can be included in environ-mental communication, using a sociological term, *societal teleconnections*, to connect local actions to specific, far-reaching impacts. I first lay out the model and sociological usage of societal teleconnections, the process of highlighting the distant, otherwise invisible, ramifications of actions close to home (Moser & Hart, 2015), and explain how technical communicators can employ societal teleconnections as a strategy to incorporate the voices of silent stakeholders in environmental communication. This involves pointing out not only the far-reaching consequences of local actions through what can feel like abstract science, but also bringing attention to the similar-ities between communities and their local knowledges, experiences, and concerns. I then present original empirical research conducted through qualitative interviews with a women's association in rural Morocco in May 2017, examining the local environmental resilience-building efforts at work by this marginalized community and connecting the localized impacts of global climate change to decisions made in the United States and other stakeholders in the Global North. In other words, I put into practice incorporating these women's narratives into the climate change debate and discussing what we learned from their local knowledge and observations, specifically how technical communicators can act as advocates. I conclude with broader implications for how TC can include and utilize the narratives of silent stakeholders to encourage environmental action.

Societal Teleconnections and Environmental Communication

Because anthropogenic climate change has such far-reaching impacts, it is important to remember that when we make environmentally irre-sponsible decisions at home, it creates a negative impact on populations and stakeholders whose voices are not typically heard or considered in the debates around how to make those decisions. Examining the distant effects of decisions made at home is a concept known as *societal telecon-nections* (Moser & Hart, 2015), a term that has been used for some time by socioenvironmental scientists to explain the ways in which environmental vulnerabilities "do not just originate and unfold in one place but can also result from long-distance relationships" (p. 14). Moser and Hart point to an often-considered example: "Several studies have examined the impacts of deforestation—due primarily to global demands for increased cropland and local slash and burn agriculture—on communities located near the sites of deforestation (Aide & Grau 2004; Lambin & Meyfroidt 2011)" (p. 15). As demand in the US grows for products such as palm oil, communities located near rainforests feel the effects of these demands from consumers across the globe. These impacts are rarely discussed in the mainstream of environmental debates. The concept of societal teleconnections can help technical communicators give a voice to distant, silent stakeholders, then, helping to foreground issues of social and environmental justice.

I recognize that many stakeholder groups in the US are marginalized as well, both as members of rural communities and as exploited workers. They also certainly deserve a voice in policymaking, as their experiences and localized knowledges are important to consider. The narratives we most often hear from these groups, however, are often constructed by corporations (e.g., coal companies and oil companies) desperate to maintain this marginalization and their monopoly on jobs and supplies. It is important for technical communicators to sort out the lived experiences of these groups from the rhetoric publicized by the corporations purporting to represent them (e.g., "the EPA kills jobs"). The simple visibility of these group—their presence in the area and their collective power to be both seen and heard—means that they are less marginalized than communities in developing nations, such as the women's association in Morocco. Yet the theory of societal teleconnections offers a way to consider both groups' experiences equally.

Local knowledge can provide important information on things such as how policy is implemented on the ground and how climate change is impacting individuals and communities. Local knowledge may be presented as local versus scientific knowledge, but societal teleconnections help communicators break down this binary as well as the binary of global versus local experiences by connecting those experiences and knowledges, by highlighting shared experiences that confirm scientific research that demonstrates global trends. For example, desertification is a known problem in Morocco amplified by anthropogenic climate change (Van Dijck et al., 2006); as the Sahara Desert creeps outward, rural areas that were once lush and green are experiencing more and more soil erosion every year (Hulme & Kelly, 1993; Scheffer et al., 2003). Meanwhile, the increased frequency and magnitude of extreme precipitation events due to anthropogenic climate change have intensified flooding and soil erosion across the Midwest (Pryor et al., 2014). Similarly, we hear accounts from farmers in the Midwest and politicians on the campaign trail about how farms are impacted by these flooding events. In the research we conducted in Morocco, and that I will describe later, we also heard clear accounts of how desertification is affecting the sheep and honey farming practices of women in this village. Using narratives and societal teleconnections to examine shared experiences and connecting those experiences to empirical data break down global versus local, and local versus scientific knowledge barriers.

A Moroccan Women's Association: A Case Study

The broad goal of the study, as originally proposed by Drs. Walton and Petrzelka, was to learn about how CSOs function, particularly CSOs designed to improve the lives of women in underresourced populations, as well as what challenges the groups face. We were especially interested in understanding how this group was able to organize and communicate internally and externally with limited or nonexistent literacy among their members. We wanted to learn what technical communication strategies the leadership found effective both for keeping the members engaged and informed and for communicating with the government to get the support they needed. We worked closely with a rural women's association, a CSO supported by a government initiative designed to improve the livelihood and education of women by jump-starting the production of sheep and honeybees. Our group included two professors, two graduate research mentors, five undergraduate students, and two local translators. Drs. Walton and Petrzelka sought and received approval from the Utah State University Internal Review Board, protocol #8414, "Civil Society Organizations in Morocco," to conduct our research.

As we learned more about this association, their activities, and their challenges, a secondary goal developed: to understand how climate change contributes to those challenges and how local knowledge could shed light on the impacts of climate change. I was generally aware that climate change was already impacting this area of the world, but it became clear that climate change had already impacted this association and that some of the women in our study were aware of it. It also became apparent, however, that these women were able to use technical communication strategies, both formal and informal, to organize and build local environmental resilience. The narratives these women shared not only led to a deeper understanding of the impacts of climate change, but they also revealed the positive impact technical communication can have when a community empowers itself through collective environmental action.

INTERVIEW METHODS

We conducted community-based research in which we immersed ourselves in the culture of the people we were working with. In doing so in a rural community, we sought to avoid what Robert Chambers (2013) calls the *urban bias*, meaning that because urban areas are more accessible and tend to contain a greater percentage of the population, they are studied more and research therefore tends to project urban results across all areas, even rural areas. In traveling to a remote location, we were able to visit an underresourced population to better understand how CSOs are attempting to improve the lives of their communities. Before arriving in the village, we spent a few days in Marrakech, one of the larger cities in Morocco, acclimating to Moroccan culture and observing what was unique and what was familiar. Once we arrived in the village, we stayed in homes with members of the women's association, eating all our meals and spending evenings and free time with our host families (or at least the women of their families). These homestays allowed us to contextualize our research findings with our own observations while the experience in the city allowed us to separate what was unique about the local situation from what was typical of Moroccan culture more broadly.

We worked in teams of four or five to conduct 30–45-minute interviews with participants, either current or former members of the association. Each team consisted of one professor, one graduate research mentor, and two or three undergraduate researchers. To maintain anonymity of the participants while tracking which team of researchers had interviewed which participant, one team named each participant alphabetically and the other named each participant numerically, so we had Participants A, B, and C being interviewed by one team at the same time that Participants 1 and 2 were being interviewed by the other team in a different room. Interviews were conducted through one of two local translators, whom we valued as important members of our team.

We did not record or take notes during the interviews so that our participants felt more comfortable about opening up to strangers and sharing their opinions, especially with regard to challenges to the association's goals.[1] Instead, each team conducted separate, simultaneous interviews, then once the participant had left, team members recorded their recollections as quickly and accurately as we could. The team would then regroup and first, compare notes with one person keeping a master set of themes, key takeaways, and any specific quotes we could recall and, second, ask the translator to confirm our understandings from the interview. Finally, both teams regrouped to discuss our separate findings. The master sets of notes were shared with all researchers at the end of the study, so responses in this chapter include those from interviews conducted by both teams as well as from my personal set of notes that might contain details not included in the master set but still corroborated by the group's short-term recollections.

LIMITATIONS

Working across languages can present limitations, most obviously that sometimes things are lost in cultural nuances, and some things just do not translate. Translators also contribute to the meaning themselves, providing much-needed cultural context, and utilizing this as part of our methodology improved our understanding of the responses (Gonzalez & Zantjer, 2015; Walton et al., 2015). Both translators we worked with are women from Morocco, so they offered insights and remained available after the interviews for us to ask clarifying questions. Working from memory, even short-term memory, also has its limitations, so our collaboration on our notes was key to making sure we had gotten the important details right. Inevitably, different points of the interview and different responses seemed important to each member of the team, so in regrouping we were able to piece together the important details more accurately than any of us could on our own.

A third limitation to the study was the number of interviews we were able to conduct during our time in the village. Between the two groups, we were able to conduct eight interviews, speaking with nine different members over a period of eight days. Because of the small number, we are not able to definitively draw broad conclusions about associations in rural Morocco, nor can we assume that these women accurately represent all 50 members of the association or that their experiences are universal. We can, however, include their narratives as evidence of the impact that climate change is having on underresourced populations and the organizations that have arisen to mitigate their lack of resources, and we can make their voices heard and consider them to be stakeholders in the climate change debate.

Findings: Incorporating the Narratives

The women in this village located in the foothills of the High Atlas Mountains formed their association in 2012 with some assistance from the government. The government of Morocco had established some incentive programs both to improve the lives of women and to improve the sustainability of agricultural practices, among other goals (Plan Maroc Vert, n.d.). One of these efforts was a grant initiative to help jump-start women's associations. If the women of a village organized and petitioned the government to help them start an association (meaning that at least one of them must be literate), they would receive some sheep (this association received five pairs of sheep) and some supplementary food for the sheep, which they would continue to receive annually. Each association has autonomy for what they do with the sheep. The women in this association chose to put their names in a lot, and their elected president drew the names of women who received the first five pairs of sheep. The women whose names were drawn took home a male and a female sheep. After those sheep produced their first lambs, there was another drawing to see which women got to take home a pair. After the second generation of lambs was produced, the women who received the first generation of lambs were able to keep any other lambs for themselves to breed or to sell at the market in the bigger town nearby and thereby earn some independent income.

This association also received three boxes of honeybees from the same government initiative. The women tended the bees and collected the honey to sell in the market. The proceeds from the honey sales went into the association's funds, and they used those funds either for buying more supplies and equipment for the honey production or saving it to build a location of their own in the village—an important step I discuss later.

There were 50 members in this association, including the leadership board, which consisted of seven members: president, vice president, secretary, assistant secretary, treasurer, and two counselors. These positions were chosen by election; each member received one vote, and they voted on almost everything. Each member paid annual dues to be a member of the association. If a woman did not have the money herself, she had to ask her husband (or father or brother if she is not married) to borrow it, and if they refused, she could not join. Participant E said that the members do not do much of anything; instead, the leadership made most of the decisions, but they kept careful notes so that if even one dirham (Moroccan currency) went missing, they would know. Several of the participants told us they gathered annually to nominate and vote on leaders, decide what to do with any money they had received from the government or honey sales, draw names and determine who got the new sheep, and discuss how to handle any other issues that came up.

GOALS OF THE ASSOCIATION

We asked each study participant the same first question: "What are the goals of the association?" The responses varied only slightly and fell into three themes.

First, an immediate goal was to build a facility where they could meet and expand their association's activities. Two of the participants mentioned that they would like to see the women begin a rug-making co-op, but they needed a large space where they could set up several looms. The more important reason for needing the facility, however, came up in nearly every interview we conducted: they needed a meeting location that they could use as their official address to apply for more grants and funding from the government. They had continued to apply for more funding to expand their association, but their applications were rejected because they did not have a location. The money they earned from the bees went toward achieving this goal.

Second, the women repeatedly told us that the overall goal of the association was to bring financial independence to its members. For some, this meant financial independence from their husbands, to be able to buy themselves new clothes or other things at the market, or to be able to send their children to school—particularly their daughters, whom many of their husbands see no need to educate. Participant 3, who was not married but was educated, continued that one of her goals for the association was to "let the women see past the curtains" of their homes. She said they were treated like machines for having babies and keeping the house, but she wanted the women to get out of the house and experience things, and earning their own money was an important step toward that goal. She led the association on excursions to the nearby city, where the women went to the market together or went on field trips or even participated in things like the International Women's March in 2017. Most of the women needed their own money to be able to go on these trips, as their husbands typically would not give them money. The association brought the women hope that they would achieve agency to improve their lives and even the lives of their daughters. Participant D told us that she did not always go on the excursions organized by the association but she did send her daughters, and Participant E told us that she was extremely proud that her daughter was taking a leadership role in the association—so proud, she said, that she sometimes bragged about it.

Improving the community was a *third* and possibly the most important goal, as nearly every participant mentioned it. The women were participating in the association not just to benefit themselves; they were participating for the good of their community—their village as a whole, the women of the village, and their families. For example, the association paid part of the cost of bringing a veterinarian to the village to check on the sheep, and this meant that everyone in the village could pay a more affordable fee to have the veterinarian look at their sheep when he came to town, whether they were members of the association or not.[2] The excursions discussed earlier were also occasionally for petitioning the government to support an improvement project for the village. Participant D said she has participated in an excursion to town to advocate for having a road paved to the village and another to ask for transportation for the children to get to school. This was an extremely important step for improving literacy rates in the village, especially among young girls, because transportation allowed for safer travel to school beyond sixth grade. This goal of being able to educate children, especially daughters, is common to many women's associations and CSOs in developing nations (Duguid & Weber, 2016).

When we asked Participant E why she had joined, she kept repeating that she wanted the association to improve the village and the community of women. We continued to ask her if there was something specific that *she* was hoping to gain from it, but she kept saying she wanted benefits for the community. When we finally asked it in a way where she understood that we meant something she wanted for herself, she said, "No, I don't have any goals for myself. That's not possible. There's no point to that. If I do something that benefits me, I want it to benefit everyone." This response reflected a sentiment we found in many participants when asked about why they joined the association: they were not doing this just to benefit themselves, though financial independence is a personal goal for many of them. The primary goal for members was to benefit everyone and to improve the livelihoods of everyone in their community, whether they were members or not.

IMPACTS OF CLIMATE CHANGE ON THE ASSOCIATION'S ACTIVITIES

The actions of others outside the village, even outside of Morocco, have already had an impact on the women's ability to achieve their goals. As discussed in the introduction to this chapter, anthropogenic climate change is projected to impact rural areas in developing nations hardest (Adger et al., 2003; IPCC, 2014, 2018, 2021). We were able to hear firsthand how this rural community had already been impacted by climate change. What follows are specific enterprises of the association and how climate change impacted them.

> *Honey production.* Climate change was most notably impact-ing the association's honey production. In a group interview with Participants A, B, and C, and in a solo interview with Participant 3, we learned that there was a drought in the area in the summer of 2016, which meant there were "no roses," or not enough pollen-producing vegetation for the bees. Because there was no food for them, the bees either had to leave the area to find vegetation elsewhere (and likely died trying) or they died. Participants 3 and A told us to prevent this from happening, the women had to feed the bees sugar to keep the insects producing honey. Participant 3 said the women disliked doing this because, for one thing, it added to the cost of honey production, and for another, it lowered the quality of the honey and therefore the price they could charge for it at the market.

> *Sheep.* Drought also impacted sheep production, as it was dif-ficult to grow the alfalfa necessary to feed the sheep between occasional shipments of food from the government. Climate change was also expediting desertification in the Sahara Desert, increasing the frequency and severity of droughts in this region as the Sahara crept into previously fertile ground (Hulme & Kelly, 1993; Scheffer et al., 2003; Van Dijck et al., 2006). Desertification is caused by the removal of vegetation,

either to create more rangeland for grazing or by overgrazing itself, but also through increased temperatures and reduced precipitation. Once this process has begun, it is extremely difficult and resource consuming to reverse. Participant 3 also observed that the temperatures had been increasingly cold in the winter and hot in the summer, and she attributed this to global climate change. Between the increased drought and these extreme temperatures, the conditions were perfect for disease to spread quickly through the village sheep. If one woman's sheep died due to disease, or if she did not have enough alfalfa from her field to feed them, then she could only hope that another woman's sheep were thriving and producing lambs, and that her name would be drawn again to receive new lambs.

While droughts are neither uncommon nor necessarily directly attributable to anthropogenic climate change, we do know that they are expected to increase in frequency and duration due to anthropogenic climate change (Dale et al., 2001; Wehner et al., 2017). In the summer of 2017, the rains returned, and honey production resumed, but with fewer bees. As climate change intensifies, the increased threat of drought puts the association's bee project in a precarious position, and the increase in extreme temperatures makes sheep production on this small scale difficult as well. The largest flock that any of the women manages was 20 sheep, so each animal was precious. As honey and sheep were the only ways these women currently had to make money for themselves, anthropogenic climate change was threatening their ability to become economically independent.

Discussion

Technical communication is vital to all the goals of the association—from communicating with the government, to getting the association formalized and funded, to maintaining informal but organized internal communication, to keeping members apprised of what is happening. However, technical communication is especially important when it comes to the goal of improving the community. One story that Participant 3 told us was about petitioning the government to grant the association olive trees for the village. She had learned about an aspect of the Plan Maroc Vert through which established organizations could ask for olive trees they could plant and maintain locally. With the collective voice of the association behind her to legitimize her request and ensuring the ministry of agriculture that the trees would be cared for, Participant 3 was successful in this petition. Because of her petition, the government allotted 500 olive trees and gave her the agency to distribute them as she saw fit. She gave each student in the school one tree to plant and care for and distributed the rest among the families of the association. These trees now provide shade, create vegetation for the honeybees, reduce soil erosion in the women's home gardens, and help clean and cool the local air and contribute (however minimally) to a global climate solution.

When Participant 3 called upon the collective agency of her women's association, she enacted technical communication for environmental action, engaging the entire community toward building resilience. She could make all their voices heard; within Morocco, these women were no longer silent stakeholders. This was the most inspiring thing we learned from our research. Technical communication empowered these women who were otherwise overlooked and ignored to improve their own lives and develop community resilience. Technical communicators in the Global North can learn much from their work. We can also continue to empower them by sharing their narratives and considering them to be stakeholders.

TECHNICAL COMMUNICATION'S GLOBAL STAKEHOLDER RESPONSIBILITY

This case study of one women's association in rural Morocco demonstrates why it is important that we consider stakeholders who might not be visible when we discuss stakeholder engagement in environmental debates in the Global North, and that we also need to broaden our notion of who constitutes a stakeholder. The concept of *societal teleconnections* can help us and our audiences better visualize the breadth of impact our local actions have on communities' livelihoods in underresourced areas. Invoking pathos alone cannot be enough to persuade an audience to change their behaviors, especially when the object of that pathos is thousands of miles away. However, connecting the narratives of these distant, silent stakeholders to more local contexts might help local stakeholders and decision makers to develop a sense of responsibility to marginalized, distant communities. By teleconnecting these communities, technical communicators can examine how decisions made by local, dominant groups are creating impacts in marginalized, distant groups—impacts that typically go unnoticed.

In technical communication, societal teleconnections may be subtle, for example, pointing out in a policy report that there are stakeholders who are not present who will potentially be affected by local decisions and actions. It may also require being more explicit, for example, including actual narrative examples from these marginalized communities when presenting research to the public. I have often encouraged STEM colleagues whose field work in environmental sciences takes place abroad to highlight local knowledges and experiences in the presentation of their work and to find ways to directly connect those experiences with the specific Western audience. For example, when I presented my work from Morocco to groups in Utah where water concerns are more dire every year, I connected this women's association to the local audience with images of the increasingly arid landscape and discussed the similarities, rather than the differences, between the local and global experiences. I discussed that, much like in communities in Utah, the family unit is keystone to the community, and that these women also have their own goals, most commonly to give their daughters opportunities they never had. I first drew connections between the culture and climate of Morocco to that of Utah, and secondly, I connected the effects of global climate change on these women's collective goals *and* the actions of the local audience.

Technical communicators may not always be able to know their audience's preexisting values, and it may often be inappropriate or uncomfortable for us in a professional context to tap into local value systems. Yet, understanding what community values and dynamics are at work can move us toward a more tailored form of communicating environmental issues that will engage our audience in a broader perspective and will break down global/local barriers.

Conclusion

Narratives like those we gathered from the women in this association in rural Morocco are important to consider in the debates surrounding environmental issues. It is well established that technical communicators have a responsibility to include stakeholders in these types of debates: we consider the coal-mining community's concerns regarding closing a coal mine, or the concerns of people who do not want a wind farm installed in their community. Yet it is also extremely important that in considering these issues, we consider the concerns of the silent stakeholders. We must seek out and listen to their narratives, write them down, carry them home, repeat them, and include them in the conversation. What impacts will the decisions we make regarding our local environment have on these women in Morocco? It may seem distant and of such little significance that it will not matter to local stakeholders, especially stakeholders who have not accepted the concept of anthropogenic climate change. But for technical communicators dedicated to social justice and engaged in environmental action, it is vital that these narratives be included.

Walton and Jones argue that "communication (written, verbal, visual, and technological) is an inextricable part of social justice because *change occurs through communicative practices*" (2013, p. 33, emphasis mine). Sarah Beth Hopton reminds us, "Technical communicators must not abuse their persuasive talents. She must not forget that people are affected by what our documents pre- and proscribe" (2013, p. 67). Technical communicators are in a unique position to include these narratives of stakeholders who otherwise have few ways of making their voices heard. Because we have that capability, we must make use of it when engaging in debates about environmental issues. When we set out to engage local stakeholders and to understand their concerns in forming arguments and in framing communication for environmental action, we have an opportunity to include these silent stakeholders. In this way, we can change the way environmental debates occur to give equal weight to the impacts on humans in developing nations, and we can begin to fight to make their voices heard.

Because global climate change is especially impacting underresourced populations who have a limited ability to communicate their own narratives to those who are adding most dramatically to the problem, it is our responsibility to recognize this is not only an environmental justice issue but more prominently a social justice one. The planet will adapt to global climate change in some way, quite likely beyond our recognition of it as natural, but it will adapt. Some human populations might even be able to adapt and survive, but those populations are the elites, the wealthy, and heavily resourced. The major impacts of anthropogenic climate change will be on rural populations in developing nations, and because most of the people living in these areas are already disenfranchised due to limited resources and low literacy rates (among other factors), technical communicators have a responsibility to bring the narratives of the silent stakeholders into the climate change conversation. This might be in the form of directly advocating for groups we have encountered in our own research and understanding how climate change has impacted their lives. It might be sharing the narratives of individuals and collectives in marginalized communities with our communities closer to home through educational outreach. It might be highlighting the global impacts of local decisions in policy discussions around environmental action. It might be pointing out the similarities between communities at home and abroad and connecting local environmental action with far-reaching consequences. Whenever we engage in technical communication for environmental action, we need to examine who is being counted as a stakeholder and make sure that silent stakeholders are heard.

Acknowledgments

I owe a very special debt of gratitude to several women who made this research possible: Dr. Aicha Lemtouni, Institute for Leadership and Communication Studies, Rabat, Morocco; Dr. Rebecca Walton and Dr. Peg Petrzelka, Utah State University; Dr. Breeanne Matheson, Utah Valley University; Tyler Oslund, Krista Larsen Black, Miranda Palmer, Emily Reeves Young, and Emily Richards, Utah State University; Hasna el Filal and Fouzya Toukart, translators. In addition, the Utah State Presidential Doctoral Research Fellowship Office and Utah State English Department provided funds that made my travel and research possible.

Notes

1. To a certain degree, this is because Morocco is largely a police state and citizens are somewhat fearful of being critical of their government. While we did not hide that we were researchers from the US and told all participants that we would use their answers for publication, our contact from the Moroccan institute informed us that the women would speak more freely about challenges they face if they did not feel that their every word was being written down or recorded for reporting to the government. We saw this as an additional barrier to their narratives being told and their voices being heard, and all the more reason for us to share their stories outside of their community. As a result, it was very difficult to transcribe very many exact quotes, though I would have loved to share their exact words in this chapter.

2. The veterinarian will also vaccinate the village dogs against rabies, which I was personally delighted to learn when one of the daughters in my host family showed me their dog's vaccination papers, and the translator explained how this worked. The dogs protect the village and the sheep from desert wolves.

References

Adger, W. N., Huq, S., Brown, K., Conway, D., & Hulme, M. (2003). Adaptation to climate change in the developing world. *Progress in development studies, 3*(3), 179–195.

Aide, T. M., & Grau, H. R. (2004). Globalization, migration, and Latin American ecosystems. *Science, 305*(5692), 1915–1916.

Bloomberg. (2017, August 10). No one is talking about this disappearing coal job. Not even Trump. *Fortune.* Retrieved from: http://fortune.com/2017/08/10/america-coal-plants-major-cuts/

Cagle, L. E., & Tillery, D. (2015). Climate change research across disciplines: The value and uses of multidisciplinary research reviews for technical communication. *Technical Communication Quarterly, 24*(2), 147–163.

Ceccarelli, L. (2011). Manufactured scientific controversy: Science, rhetoric, and public debate. *Rhetoric and Public Affairs, 14*(2), 195–228.

Chambers, R. (2014). *Rural development: Putting the last first.* Routledge.

Colman, Z., & Adragna, A. (2021, April 18). Biden takes on Dems' "Mission Impossible": Revitalizing coal country. *Politico.* Retrieved from https://www.politico.com/news/2021/04/18/coal-country-revitalization-biden-482659

Dale, V. H., Joyce, L. A., McNulty, S., Neilson, R. P., Ayres, M. P., Flannigan, M. D., Hanson, P. J., Irland, L. C., Lugo, A. E., Peterson, C. J., Simberloff, D. (2001). Climate change and forest disturbances: Climate change can affect forests by altering the frequency, intensity, duration, and timing of fire, drought, introduced species, insect and pathogen outbreaks, hurricanes, windstorms, ice storms, or landslides. *BioScience, 51*(9), 723–734.

Dodson, G., & Palliser, A. (2016). Advancing practical theory in environmental communication: A phronetic analysis of environmental communication in New Zealand. Iowa State Symposium on Science Communication.

Druschke, C. G., & McGreavy, B. (2016). Why rhetoric matters for ecology. *Frontiers in Ecology and the Environment, 14*(1), 46–52.

Duguid, F., & Weber, N. (2016). Gender equality and women's empowerment in co-operatives. *A literature review.* International Co-operative Alliance.

Fahey, D. W., Doherty, S. J., Hibbard, K. A., Romanou, A., & Taylor, P. C. (2017). Physical drivers of climate change. In D. J. Wuebbles, D. W. Fahey, K. A. Hibbard, D. J. Dokken, B. C. Stewart, and T. K. Maycock (Eds.), *Climate science special report: Fourth national climate assessment, volume 1* (pp. 73–113). US Global Change Research Program. doi:10.7930/J0513WCR

Gonzalez, L., & Zantjer, R. (2015). Translation as a user-localization practice. *Technical Communication, 62*(4), 271–284.

Hood, G. (2018, August 21). Debate brews in Pueblo over the balance of coal and renewable energy. *NPR: Morning Edition.* Retrieved from https://www.npr.org/2018/08/21/640437958/debate-brews-in-pueblo-over-the-balance-of-coal-and-renewable-energy

Hopton, S. B. (2013). If not me, who? Encouraging critical and ethical praxis in technical communication. *Connexions: International Professional Communication Journal, (1)*1, 65–68.

Howe, P., Mildenberger, M., Marlon, J., & Leiserowitz, A. (2015). Geographic variation in opinions on climate change at state and local scales in the USA. *Nature Climate Change.* doi:10.1038/nclimate2583

Hulme, M., & Kelly, M. (1993). Exploring the links between desertification and climate change. *Environment: Science and Policy for Sustainable Development, 35*(6), 4–45.

Intergovernmental Panel on Climate Change (IPCC). (2014). *Climate change 2014—impacts, adaptation and vulnerability: Regional aspects.* Cambridge University Press.

Intergovernmental Panel on Climate Change (IPCC). (2018). *Global warming of 1.5°C: An IPCC special report on the impacts of global warming of 1.5°C above pre-industrial levels and related global greenhouse gas emission pathways, in the context of strengthening the global response to the threat of climate change, sustainable development, and efforts to eradicate poverty.* Intergovernmental Panel on Climate Change.

Intergovernmental Panel on Climate Change (IPCC). (2021). *Climate change 2021: The physical science basis.* Contribution of Working Group I to the Sixth Assessment Report of the Intergovernmental Panel on Climate Change. Cambridge University Press.

Jones, N. N., Walton, R., Haas, A. M., & Eble, M. F. (2018). Using narratives to foster critical thinking about diversity and social justice. In A. Haas & M. Eble (Eds.), *Key theoretical frameworks: Teaching technical communication in the twenty-first century* (pp. 241–267). Utah State University Press.

Lambin, E. F., & Meyfroidt, P. (2011). Global land use change, economic globalization, and the looming land scarcity. *Proceedings of the National Academy of Sciences, 108*(9), 3465–3472

Moser, S. C., & Hart, J. A. F. (2015). The long arm of climate change: Societal teleconnections and the future of climate change impacts studies. *Climatic Change, 129*(1–2), 13–26.

Plan Maroc Vert. (n.d.) *Maroc.ma.* Retrieved from http://www.maroc.ma/fr/content/plan-maroc-vert

Pryor, S. C., Scavia, D., Downer, C., Gaden, M., Iverson, L., Nordstrom, R., Patz, J., & Robertson, G. P. (2014). Midwest. In J. M. Melillo, T. C. Richmond, & G. W. Yohe (Eds.), *Climate change impacts in the United States: The third national climate assessment* (chap. 18, pp. 418–440). US Global Change Research Program. doi:10.7930/J0J1012N

Querol, X., Alastuey, A., Lopez-Soler, A., Mantilla, E., & Plana, F. (1996). Mineral composition of atmospheric particulates around a large coal-fired power station. *Atmospheric Environment, 30*(21), 3557–3572.

Scheffer, M., Carpenter, S. R., Foley, J. A., & Walker, B. (2003). Catastrophic regime shifts in ecosystems: Linking theory to observation. *Trends in Ecology and Evolution, 18*(12), 648–656.

Simmons, W. M. (2008). *Participation and power: Civic discourse in environmental policy decisions.* State University of New York Press.

Simmons, W. M., & Grabill, J. T. (2007). Toward a civic rhetoric for technologically and scientifically complex places: Invention, performance, and participation. *College Composition and Communication,* 419–448.

Van Dijck, S. J., Laouina, A., Carvalho, A. V., Loos, S., Schipper, A. M., Van der Kwast, H., Nafaa, R., Antari, M., Rocha, A., Borrego, C., & Ritsema, C. J. (2006). Desertification in northern Morocco due to effects of climate change on groundwater recharge. In *Desertification in the Mediterranean Region: A security issue* (pp. 549–577). Springer.

Walker, G. B. (2007). Public participation as participatory communication in environmental policy decision-making: From concepts to structured conversations. *Environmental Communication, 1*(1), 99–110.

Walton, R., & Jones, N. N. (2013). Navigating increasingly cross-cultural, cross-disciplinary, and cross-organizational contexts to support social justice. *Communication Design Quarterly, 1*(4), 31–35.

Walton, R., Zraly, M., & Mugengana, J. P. (2015). Values and validity: Navigating messiness in a community-based research project in Rwanda. *Technical Communication Quarterly, 24*(1), 45–69.

Wehner, M. F., Arnold, J. R., Knutson, T., Kunkel, K. E., & LeGrande, A. N. (2017). Droughts, floods, and wildfires. In D. J. Wuebbles, D. W. Fahey, K. A. Hibbard, D. J. Dokken, B. C. Stewart, and T. K. Maycock (Eds.), *Climate science special report: Fourth national climate assessment, volume 1* (pp. 231–256). US Global Change Research Program. doi:10.7930/J0CJ8BNN

Chapter 7

Resilient Farmland

The Role of Technical Communicators

SARA B. PARKS AND LEE S. TESDELL

When we first began discussing the practical applications of technical communication in agriculture resilience and environmental action in the US Midwest, it was 2017 and we had both watched three years of failed litigation between the Des Moines, Iowa, Water Works and Iowa Drainage Districts dominate the local news. With the Mní Wičóni Movement of Standing Rock versus the Dakota Access Pipeline as well as the continuing national scandal of water contamination in Flint, Michigan, driving the conversation about clean water in US national news, we watched outrage and hope be gradually tempered by political realities and inequitable systems. Our concern then, the problem that is reflected in this chapter, was how technical communication and technical communicators can continue to engage in environmental action even when the *kairos* seems unamenable to policy change.

Now writing in 2021, much has changed. We are tentatively hopeful that intersectional environmentalism has found a foothold within the momentum of broader social movements. We are rethinking how to engage and hold outreach events under COVID-19 social distancing restrictions that might translate to broader access in the future. We are again hopeful that we might see positive change in national environmental policy. Yet the basic situation, the problem of *kairos* for technical communicators who seek to engage with US midwestern agricultural practices, remains much the same.

In this chapter, we argue that the best solution for helping practitioners engage in situations with uncertain or poor *kairos* is teaching practitioners to better read and navigate context. Therefore, we begin this chapter by explaining the context for technical communication in the $1.1 trillion agriculture industry (USDA, 2017) as well as reviewing some typical communication roles and challenges in this context. Then we present a couple of cases in the Midwest where technical communication-supported decision making is helping to safeguard our agricultural land by informing and encouraging communication between decision makers, publics, and experts about the ravages of climate change and agricultural innovations that might mitigate them. We end with potential solutions by detailing ways that practitioners might continue to engage in promoting resilient agriculture.

Knowing a Place to Understand a Case

The first step for a technical communicator to better understand context is to read about the place's historical context. The Midwest is a large area of the United States with relatively undefined boundaries. One of the challenges of applying technical communication to midwestern agriculture is gaining a thorough understanding of the local place. For example, the Flint Hills of Kansas are a very different landscape than the Ozarks of Missouri or the Black Hills of South Dakota, even though all are considered "hilly" areas in a region most known for its (flat) Great Plains. Agriculture techniques, timing, risks, and opportunities all differ by local area. For better focus, the cases we review in this chapter are in Iowa, particularly the "Prairie Pothole," central and northern regions of Iowa. We begin with an overview of Iowa's agricultural history and its unique agricultural challenges to demonstrate the breadth of knowledge technical communicators should have about their chosen place.

The state of Iowa has about 23 million acres of corn and soybeans. This row-crop ground dominates Iowa's landscape today. However, when European immigrants came to Iowa, they found the southern third of the state hilly and wooded while the northern two-thirds was covered largely by tall-grass prairie and wetlands with wooded areas along the rivers. The northeast corner of the state was hilly and forested as well. The territory of Iowa was ceded by native people in a series of treaties with the Sac and Fox (in 1824, 1830, 1832, 1836, 1837, and 1842), the Potawatomi (1846), and Sioux (in 1853) (Sherman, 2015). There is archeological evidence of cultivation in Iowa of little barley, an annual grass, as well as other native plants up to 3,000 years ago (Office of the State, 2020). However the major change in Iowa's landscape, draining the marshes and wetlands of its Prairie Pothole region, was due to a focused pursuit of agriculture as a commodity by European settlers.

Over time, Iowa farming turned to cash commodities and concentrated on corn and soybeans ("rowcrop agriculture") as well as eggs and swine ("protein agriculture"). Corn and soybeans have continued to be the main crops of Iowa in large part due to Iowa's rich, fertile topsoil that, until recently, has seemed able to withstand the nutrient-depleting and eroding impact of monoculture (Eller, 2014). To raise large amounts of these two grains, the former prairie was nearly all plowed up and the wetlands nearly all drained; with these two major changes to the landscape, Iowa farmers created a "leaky" agricultural system that drains excess nutrients from agriculture into its waterways.

Three practices led to excess nutrients in the watersheds of Iowa. First, annual crops replaced perennial vegetation. Second, frequent tillage has damaged the soil structure of Iowa fields. In Iowa, farmers often till fields in the fall so that the dark earth revealed will warm faster in the spring to allow earlier planting. The problems with this practice are that tillage results in 1) soil erosion, 2) a sealed "crust" on the soil's surface that leads to more surface runoff, 3) lack of plant matter to filter surface water, and ultimately 4) loss of topsoil and 5) poor soil quality, which can keep roots from growing deep, which affects yield (Al-Kaisi, 2020). Third, the drainage tiles that now honeycomb Iowa's landscape allow excess nutrients to easily escape into the watersheds. Drainage tiles are pipes that allow excess water to be drained from flat areas of swamp and marshy land. Initially, the "pipes" were clay tiles placed end-to-end with a small gap to allow water to drain out of the soil. Today they are made of perforated plastic. There are no complete records of how much drainage has been installed in Iowa. However, the USDA estimates Iowa holds a quarter of all drain-tiled acres in the US, an estimated 14 million acres with an estimated 800,000 miles of drainage tiles (USDA, 2017; Love, 2010). That's enough pipe to wrap around the earth 32 times. Drainage tiles allow farmers to grow row crops on high-quality soil that would be otherwise unfarmable.

While nitrate occurs naturally in the soils of Iowa, Iowa farmers add 150–200 lbs. per acre to most corn fields to increase the yield. The plants do not use all this nitrate and so heavy rainfalls flush it out of the soil through the drainage tiles and into the watersheds and rivers. The excess nitrate and other excess nutrients pollute the water downstream. Herein lies the environmental, social, political, and economic problems the cases discussed in this chapter take on.

Rhetorical Network for Technical Communication in Agriculture

The second step for a technical communicator to better read context is to identify the main actors who create the unique rhetorical network that the technical communicator wishes to engage. Technical communication has unique rhetorical features in the US midwestern agricultural context. For example, landowners and farmers are currently the main actors whose decisions make farmland resilient to climate change. In contrast to environmental actions in other arenas, the technical communicator who wishes to promote resilient agriculture will most likely accept landowners and farm managers, not policymakers, as their main audience or users. In times of questionable *kairos*, the challenge for technical communicators is deciding how to position themselves (and then positioning themselves) in this rhetorical network.

Environmental action groups that engage in US midwestern agriculture are often steeped in academic frameworks and/or staffed with academics and researchers. This is likely due to long-standing partnerships with Cooperative Extension programs and other government/academic partnerships and support. Therefore, it is important for the technical communicator who will likely be supported by these networks to understand how academics generally view environmental action. In academic fields that concern the environment of privately owned land, such as grain production agriculture, forestry, wetland management, wind energy, and bioenergy, much attention is paid to individual landowner and farm manager attitudes and perspectives. Because conservation and other government programs are usually voluntary, these programs must strategize their marketing and create frameworks for evaluation from case-based, industry-specific, and regional research about their potential participants (Haggerty et al., 2018; Stroman & Kreuter, 2016; Swinton et al., 2017).

In this literature, the single landowner is often framed as a problem: an unwilling or skeptical but central research subject, a blank slate who lacks knowledge about climate science or conservation programs, a legal entity with unfortunate land usage rights (Cocklin et al., 2007; Eaton et al., 2007; Tulloch et al. 2014). The technical communicator who becomes involved in environmental action groups could easily adopt these deficit model attitudes and critical academic focus on key individuals.

However, the US midwestern agricultural system itself provides a check on this attitude and focus. The context is a more complex system of rhetorical influences than reflected in the landowner attitude literature. A quick analysis reveals hidden agents such as drainage tile contractors, fertilizer suppliers, equipment dealers, spray rig operators, seed suppliers, sales agronomists, bankers, farm managers, and even the retired farmers who meet for coffee and gossip at the local cooperative office every morning. Each of these people has an opinion that influences decision making. Often these people are also landowners and farm managers' friends, relatives, and neighbors. Their opinions matter in decision making but they can be overlooked.

The persistence of the family farm mythos also complicates a technical communicator's understanding of the rhetorical context. The family farm does exist—in fact, the authors of this chapter have both lived on multigeneration family-owned, small midwestern farms. However, according to a 2017 Farmland Ownership Tenure Survey conducted at Iowa State University, the family farm is not as pervasive as it might seem in the Midwest. In Iowa, 90% of the land is farmland, however, 60% of Iowa's farmland owners don't actively farm. In fact, 34% of these owners have no farming experience at all. Instead, 53% of Iowa's farmland is leased. Who actually owns Iowa's farmland is also often misrepresented. According to the ISU survey, 60% is owned by people aged 65 or older. And women own 47% of Iowa farmland (Zhang et al., 2018). The family farm mythos of a patriarchal family where young to middle-aged parents are willing to assume the risk of innovation to make a better world for their children no longer reflects most midwestern agriculture's decision makers.

Beyond the obvious problems for a technical communicator in correctly identifying their potential users or audience, the family farm mythos also can contribute to agricultural decision makers and environmental sustainability advocates misunderstanding each other. For example, this can happen through a stasis definition mismatch for the term *sustainability*. In the family farm mythos, sustainability can mean that a family retains their land and farming traditions and passes them to the next generation. Sustainability can lead to risk-aversion because a farm must stay solvent so the family can keep the land. Sustainability sets up roadblocks to innovation because farming traditions are codified as family traditions. Therefore, when a scientist claims row-crop agriculture, for example, is not sustainable, a midwestern farmer might think back on their life and say, "Well, it sustained us just fine!" (This is also why the authors of this chapter prefer the term *resilience* for its forward-looking connotations.) Even though the family farm is a mythos in rhetoric, the values that stem from it still influence decision makers' identities and thought processes.

We should also recognize that farmers are not monolithic in their attitudes toward adopting new practices on the land. In fact, recent research proposes a typology of four types of farmers: Conservationist, Deliberative, Productivist, and Traditionalist (Upadhaya et al., 2020). A technical communicator should continually assess and reflect on the impact of different types of farmer attitudes and, perhaps just as important, what types of landowners are present and the attitudes they might hold. For example, in turn-of-the-century Iowa corn farmers were persuaded to use hybrid corn seed. But this adoption didn't happen overnight. Biographers to US vice president and one of the founders of Pioneer Hybrid Seed Company (now DowDupont), Henry A. Wallace, noted, "Often it took three or four years to persuade a farmer that the large yields produced by hybrid seed were not a fluke" (Culver & Hyde, 2000, p. 91). Wallace believed two important principles would succeed in marketing the new hybrid seed to farmers: total honesty and high prices. High prices, Wallace believed, would convince farmers "they were buying something special" (Culver & Hyde, 2000, p. 91fn). This example is just another case that shows agricultural decision making (like most human decision making) deviates from pure economic logic. Technical communicators must understand the rhetorical sensibility of the network and their audience or users.

Navigating the agriculture communication network is likely the trickiest, but most important, practice a technical communicator who wants to promote resilience can do. Technical communicators' rhetorical skills are uniquely beneficial to this context. Plant science researchers Streit Krug and Tesdell acknowledge the need for a more holistic perspective to encourage perennial grain agriculture: "Recognizing the dominant norms, values, and patterns of relationship in the agricultures that currently feed most people—that is, beginning to gain awareness of how they influence human ways of seeing, thinking, feeling, and knowing, so that some things are more or less visible, thinkable, sensible, and knowable—matters because these same norms, values, and patterns shape the very cultural tools and social institutions available to try and enact positive agroecological change" (Streit Krug & Tesdell, 2020). However, awareness needs to translate into action. This is the step at which technical communicators can stand up and do practical work.

Common Genres of Technical Communication in US Midwest Agriculture

The third step for a technical communicator to better read context is to learn and assess the benefits and challenges of the standard genres used in that context. There are a few standard genres employed in the midwestern agriculture industry. These genres may no longer be the best ways to communicate, and they are likely undergoing genre change due to our current pandemic situation and technology capabilities. However, they are also genres that are not as prevalent in other industries, so it is important to identify and acknowledge their common use in agriculture outreach. Genres such as the field day, the radio talk show, maps, and nonmarketing industry publications are all important. We will cover field days and maps in this chapter since both are genres an entry-level technical communicator would be able to most easily access and influence.

The Agriculture Field Day

Agriculture field days are site visits that invite farmers, landowners, and other key actors in the agriculture system, such as managers, consultants, and agents, to the (usually literal) field in order to see systems, tools, and new techniques in action. Formalized field day events have a very long history in agriculture innovation.

Field days were and continue to be important during international dignitary visits. For example, Pezzullo and Hunt identify the field day visit to an Iowa farm that Soviet premier Nikita Khrushchev made in 1959 as pivotal in American, Russian, and Chinese economic relationships in the decade that followed (Pezzullo & Hunt, 2020). Field days continue to be regularly held to educate new policymakers and governmental staffers and to publicize new systems, techniques, and tools.

The most common field day structure is a mix of guided tour (walking through sites to observe the impact of tools and processes) and listening to presentations at a home base area, often in a barn or tent set up near the field site. Accessibility is often a concern, especially as the population of landowners and farmers has aged. Lunch is often included in the event. Most often, field days are small social events, involving 20–60 people total, and they leave plenty of time for informal Q&A, networking, "kicking the tires" of new machinery, and shop talk.

The expert content of field days varies widely. Michael Carolan's (2008) thorough exploration of agricultural field days in "Democratizing Knowledge" compares conventional agriculture's field days to sustainable agriculture's field days. He found that while conventional agriculture's field days adhere to conventional assumptions "about values, decision stakes, social purposes, and the like," which work to establish conventional designations about who is an expert in the field day (and therefore, whose words and worldviews have the most weight during knowledge coproduction), sustainable agriculture's field days are more likely to explicitly discuss "what and who the goals, stakes, and experts are" (p. 514). Essentially, sustainable agriculture's field days are more open to big-picture questions that move beyond the assumptions and goals of agriculture that has profit as its only motive. Carolan notes that at sustainable agriculture field days, the expert who should respond is "tightly coupled" to the question asked (p. 522).

However, this critical awareness can also create challenges since one of Carolan's participants notes, "We can all talk ourselves to death" (p. 524). Sustainability agriculture field days have significant challenges in identifying when to end deliberation and move to collective decision making. This is a reminder that a successful event should have a mechanism for participants to reflect, assess, and make concrete, future plans. It is an area that a technical communicator can positively impact by providing outlines, rubrics, good reflective questions, and other communicative support.

Events such as field days fall into the "Broader Impacts" category for assessment by the National Science Foundation. While not all field days are sponsored by science organizations such as the NSF, many are cosupported in some way or "leveraged" by projects that need public events to fulfill outreach criteria for their funding. The NSF's (2020) broader impacts category is much debated and criteria for it are often revised. However, the impact assessment suggested for broader impacts events captures the challenges of assessing agriculture field days.

In the NSF Merit Review FAQs for broader impacts, assessment is flattened to "document the results" and "document the outputs of those activities" as if broader impacts events take place in a lab where clear-cut documentation is possible within the time scope of the project

(NSF, 2020). In practice, assessment often ends at reporting the size of audiences. Better assessments take note of demographic information and the role of audience members (whether they are policy makers, experts, producers, etc.), and the best examples of event assessment successfully administer a post-test or gather reflections for qualitative evidence of impact. Very few projects can implement long-term impact analysis for various reasons, not the least of which is that most projects are funded for five years only.

Getting the timing right for assessment to make reasonable conclusions about event impact has also been identified as a challenge in Kamau et al.'s (2018) assessment of the impact of demos and field days in sub-Saharan Africa. The researchers studied the impact of augmenting radio promotions for an improved corn seed with demonstration plots and field days. They found that while there was a slight increase in awareness and positive perception from the group of farmers who had access to demonstration plots and field days, that awareness did not translate to more use of the new seed. The authors' limitations statement captures the confounding factors of assessing field days. "The findings may have been influenced by the constraints or challenges faced by the seed company in rolling out the promotion campaign as agreed. These constraints include delays in rainfall, inadequate rain, destruction of demo plots by livestock, poor or incorrect labelling and destruction of sign boards at demo, unavailability of the demo hosts at their plots to answer other farmers' queries about the new varieties and poor turnout during field days" (Kamau et al., 2018). While lack of rain and wandering livestock are outside a technical communicator's control, supporting signage, strategizing timing, as well as collecting and interpreting usability assessment are areas in which a technical communicator could contribute.

Maps

Another common method of conveying information in the agriculture context is maps. Maps are ubiquitous. They show physical trends throughout a state, a county, or one's own land. Mapping in agriculture is currently an area of high innovation, tied up in geographic information systems (GIS) and remote sensing possibilities, as well as current integration in agricultural technology automation. It is likely mapping will become even more interactive, granular, and expected in technical communication. However, the high-tech possibilities do come with challenges.

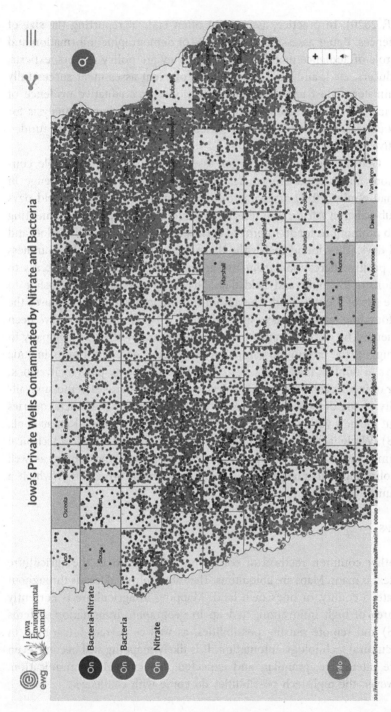

Figure 7.1. Iowa private wells contaminated by nitrate and bacteria map.

Figure 7.1 shows an interactive map published by the Environmental Working Group and Iowa Environmental Council (2019). The map shows where private wells are contaminated in Iowa. The data is from a database maintained by the Iowa Department of Natural Resources from 2002 to 2017 with two years of data thrown out for likely inaccuracies. The map loads with all pollutants toggled "on" for high visual impact. Zoom capabilities allow granular exploration. The map is supported by an article with more charts, clear explanation of the limitations, and methods. This is an excellent piece of technical communication.

However, as will be discussed later, during the exploration of the Des Moines Water Works case, technical communication in agriculture tends to "borrow" well-designed images and repurpose them. In rural Iowa, where cell signal is spotty and wi-fi doesn't reach past the lathe-and-plaster living room, how will this map be presented? What will happen to this map when it is printed in black and white and handed out during a farm tour or rural workshop? What happens when the limitations of the map and its methodology are separated from the image? Figuring out how to create technical communication that can be remixed, repurposed, and reused without significantly declining quality is a constant challenge for technical communication in the agriculture context.

The importance of classic technical documents, such as maps, and challenges of handling communication that brings together the rhetorical network, such as field days, is exemplified in the next two cases. The Des Moines Water Works Case is a negative case. Although we believe technical communication has little to do with the outcome of the case, it shows how the complicated agricultural rhetorical network circulates communication and suggests opportunities for technical communicators to make a positive impact. Our second case, the Prairie STRIPS case, is a positive example of technical communication's impact on resilient agriculture. Both cases show how opportunities for engagement can still exist in situations with uncertain or poor *kairos*.

Failed Action: The Des Moines Water Works Case

Des Moines Water Works started in 1871 in Des Moines, Iowa, as the Des Moines Water Company. Today it draws water mostly from Raccoon River watershed and supplies about 500,000 customers. Des Moines Water Works is legally responsible for removing pollutants and excess nutrients to make the water safe for its customers to use in their homes (Des Moines Water Works, 2020). In 2015 the Water Works board decided to sue rural drainage districts due to the pollutants that were flushing downstream.

BACKGROUND TO THE LAWSUIT

In a move to assign responsibility for pollutants (mostly excess nitrates) in the water, the Des Moines Water Works board sued three counties' boards of supervisors that are trustees for 10 drainage districts in Northwest Iowa. The Water Works contended that because pollutants originated in northwestern Iowa's farm fields, the Water Works had to install and operate expensive equipment to bring the nitrate level to below 10 mg/l, the federal mandate for drinking water (Des Moines Water Works, 2020).

The lawsuit made waves across national agriculture publications and played out in a series of articles and blistering letters to the editor in the main Des Moines newspaper, the *Des Moines Register*. Technical communication, largely in the form of watershed maps, pollutant charts and graphs, and technical descriptions of water testing equipment and procedures dominated the public discourse. While this lawsuit was eventually tossed out of court, the suit suggests that there is support, even in such a conventional agriculture-friendly state as Iowa, for public accountability and the involvement of public, even urban, voices in agriculture management decision making. These voices have been missing from agriculture discourse.

WHY THE LAWSUIT MATTERS

The Des Moines Water Works case is a very public and dramatic attempt at accountability. In Iowa, it serves as a first attempt and a model for more lawsuits. A new lawsuit filed by Iowa Citizens for Community Improvement and Food and Water Watch is currently (in 2021) being heard at the Iowa Supreme Court. Again, playing out in public via the *Des Moines Register*, the lawsuit targets Iowa's hog farms as well as the voluntary nature of farm pollution controls. It explicitly ties agricultural practices to unsafe drinking water for Des Moines. For example, AP reporter David Pitt summarized the lawsuit as asking "whether enough is being done to keep hog manure and other farm pollutants from tainting rivers that provide central Iowans drinking water" (Pitt, 2019). That a precedent might be set with these lawsuits is acknowledged by the lawyers for the Iowa Attorney General's Office who argued to dismiss the case because, "a court ruling that would impose nitrogen and phosphorous restrictions on farms 'would be a first in the nation and a dramatic shift from present-day agricultural practices'" (Pitt, 2019). If lawsuits such as these continue in Iowa and spread to other states that are dependent on agriculture, the Des Moines Water Works case may end up symbolizing a dramatic change in who can influence agriculture policy.

THE TECHNICAL COMMUNICATION OF THE DES MOINES WATER WORKS CASE

Public technical communication during the Des Moines Water Works case of 2015–2017 stemmed from Water Works data and was supported by various communicators and educators in university extension and environmental groups. For example, figure 7.2 shows a map of Iowa that was often reprinted during the lawsuit both in color and in grayscale.

The source of the map is Kaye LaFond from the global pro-water data, design, and journalism company Circle of Blue (2015). The map was published in March 2015 as part of a Circle of Blue article introducing the

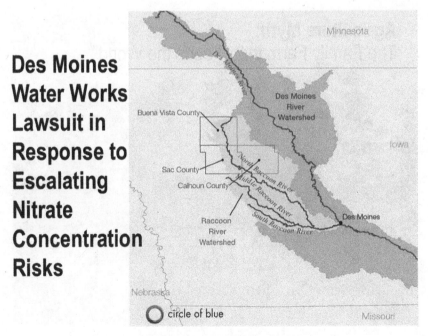

Figure 7.2. Circle of Blue map.

water works lawsuit. The image appearing in figure 7.2 is clipped from a slide set created by Water Works CEO Bill Stowe (n.d.) or his staff to introduce the Water Works lawsuit. This linking of sources, images, and organizations is typical of the Water Works' public technical communication. Borrowing (without academic citation, but usually with attribution or a logo) is typical of public technical communication in agriculture as well, in our experience.

The Circle of Blue map shows the Raccoon River and its watershed through blue lines and shading. It also shows Buena Vista, Sac, and Calhoun counties, but not their water districts, which have slightly different boundaries, and the map includes the location of Des Moines at the junction of the Des Moines and Raccoon Rivers. However, the font size and pin size make Des Moines easy to miss. The map works very well in its original setting—an online article with the option to "click to enlarge." However, its reprint in a variety of genres and modes didn't always work so well.

The technical communication from the Water Works and other lawsuit-supporting groups varies in quality. In another slide from the same Water Works set (fig. 7.3), the statistic that fulfills the argument is buried by poor design.

Agriculture Myth:
The Family Farm that "Feeds the World"

Approximately 90,000 Iowa farms
Nearly 90% of Iowa's food is imported

Figure 7.3. Water Works slide: agriculture myth.

It is easy to give examples from this one poorly designed, but important, Water Works slide set. But what about communication from other organizations? One of the activist organizations that supported the lawsuit and became heavily involved in public discourse was the Iowa Citizens for Community Improvement (CCI). The Iowa CCI has three current campaigns—"Farm & Environment," "Fight for $15," and "Winning Racial Justice." These show the organization's broad spectrum as well as integration of social movements. This is the organization currently spearheading the Clean Water lawsuit against the State of Iowa.

Iowa CCI members wrote letters to the *Register* and attended events and meetings regarding the Water Works lawsuit. Their website was, and still is, primarily text based. A screenshot appears in figure 7.4.

The website is highly functional for one-way communication such as events announcements and reprinting letters (such as the one in the image that obliquely references the Water Works debate). More engagement appears on their social media. Their Facebook page has over 10,000 followers. However, posts are still mostly reposts of letters and articles, primarily written opinions and primary-source research, completed by CCI members and organizations such as the University of Iowa's Hydroscience and Engineering blog (https://www.iihr.uiowa.edu/). And while sources like the IIHR blog are excellent expert-to-expert, informal technical communication (often with excellent maps and charts), they still assume a level of expertise beyond the public audience.

Figure 7.4. Iowa Citizens for Community Improvement website.

Groups opposed to the lawsuit garnered public support through communication such as TV commercials and articles in the *Des Moines Register*. Groups including Agriculture's Clean Water Alliance (an association of 12 agriculture retailers and supported by the Iowa Soybean Association) argued for compromise and voluntary clean water strategies in TV ads, social media, events, and tours. Presenting themselves as the reasonable, educated voice in the debate about whether lawsuits ought to impact agriculture was likely made easier with the foil of Water Works' CEO Bill Stowe capturing national headlines.

The four panels shown in Figure 7.5 are from the ACWA's social media campaign in 2016 during the "One Water Summit," a national meeting hosted by the US Water Alliance. According to ACWA chairman Harry Ahrenholtz, the ACWA sent the only representative for agriculture to the summit, which created a US Water Alliance roadmap (ACWA, 2016a).

The images appear on Facebook with the caption "public-private & urban-rural partnership—vital to enable nutrient reduction nationwide" (ACWA, 2016b) It includes the hashtag #OneWater16, used by the One Water Summit. These images seem innocuous, showing a row crop framed by farm buildings and urban buildings in the background as well as positive slogans that appeal to "Iowa nice," and compromise. However, the ACWA's public messages were debated and criticized for lack of data and misrepresenting the nature of the Water Works lawsuit (Harden, n.d.).

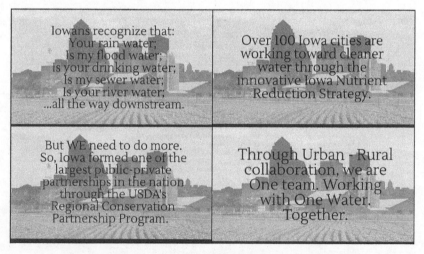

Figure 7.5. ACWA Ag Clean Water Alliance social media.

Another anti-lawsuit group, the Iowa Partnership for Clean Water (supported by the Iowa Farm Bureau), ran a TV spot in 2015 with the tagline "lawsuits aren't the answer and farmers aren't the enemy, they are the solution" (Redwave Digital, 2015) The organization (which has not been active on social media since 2018) also made a more pointed character assassination video of Water Works CEO Bill Stowe, which can still be found on Facebook along with a record of its relatively limited reach (under 300 views as of March 2021) and criticism in the comments.

In the end, the public debate and technical communication embedded in that debate had little to do with the outcome of the Water Works lawsuit—which essentially determined that water districts couldn't be held responsible for water quality and couldn't be sued. However, how the case played out in public garnered support for the use of lawsuits, or the threat of lawsuits, even while the verdict itself created a roadblock for a lawsuit's success. It shows an opportunity for more technical communication support in the push for resilient agriculture as well as the clear divide between two opposing arguments in agriculture rhetoric: that farmers are reasonable people who will voluntarily switch to practices that save our waters and soil from disaster versus a more proactive argument for resilience that also tends to be more cynical about changing our ways in time to avert climate change and further degrading our water and soil.

Successful Action: The Prairie STRIPS Case

A rhetorical situation where policy change seems both a monumental task and entirely too incremental can be very discouraging for technical communicators who support environmental action. However, the grassroots pursuit of resilient agriculture in Iowa continues and does boast successes, one of which is a popular conservation project that was begun only 35 miles away from the Des Moines Water Works headquarters.

BACKGROUND TO PRAIRIE STRIPS

Science-Based Trials of Rowcrops Integrated with Prairie Strips (STRIPS) is a conservation project that has been based at Iowa State University in Ames, Iowa, for the last decade. Dr. Lisa Schulte-Moore (Iowa State professor in Natural Resource Ecology and Management) is a cofounder. The concept of the project is to encourage farmers and landowners to cultivate native plants at the edges of row-cropped fields or plant strips of native plants within those fields. These "prairie strips" offer soil health, water quality, and biodiversity benefits.

The scientific research that led to the STRIPS project began in 2003 as a collaboration with Neal Smith National Wildlife Refuge, located in Prairie City, about 25 miles away from Des Moines. There, Iowa State researchers worked with refuge managers to integrate restored prairie into row-crop fields. A series of scientific publications document and analyze the environmental, social, and economic benefits found in these trials. The project's most persuasive numerical finding is that "by converting 10% of a crop-field to diverse, native perennial vegetation, farmers and landowners can reduce sediment movement off their field by 95 percent and total phosphorous and nitrogen lost through runoff by 90 and 95 percent, respectively" (STRIPS, 2021). Now, the project's scientific research continues at some commercial farm fields in the Midwest (STRIPS, 2021). Meanwhile, the practice of prairie strip implementation is informed and publicized by the project's myriad public technical communication documents and the organization's presence at events.

Why the Program Matters

The STRIPS project has seen clear success in interest, awareness, and implementation of prairie strips. The 2020 Iowa Farm and Rural Life Poll documented the expansion in awareness and interest in prairie strips since 2018. The report states:

> About two-thirds (66%) of farmers reported that they had heard about the practice before reading the description, up from 56% in 2018. . . . A second question asked respondents if they would be interested in learning more about the practice. In 2020 27% selected "yes" and 26% selected "maybe," compared to 22% and 36%, respectively, in 2018. Similarly, in 2020 20% responded that they would be interested in planting prairie strips on their land, and 31% indicated that they might be interested, compared to 15% and 39% in 2018. (Arbuckle, 2021)

Importantly, planting prairie strips has become an accepted and encouraged resilient agriculture practice at national levels. The USDA's Conservation Reserve Program (CRP) now supports the planting of prairie strips.

According to personal communication with STRIPS agricultural specialist Tim Youngquist, "In most cases, the annual rental payment provided to the landowner by the CRP CP43 meets or exceeds the annual rental payment they were receiving from their tenant." This national acceptance and financial support influences farmer and landowner decision making, as indicated in the poll: "To gauge farmer interest in this new option, the survey posed the question, 'Prairie strips are now eligible for annual rental payments through the Conservation Reserve Program (CRP). Would CRP payments increase your interest in establishing prairie strips?' Almost half (47%) indicated that CRP payments would increase their interest, and 22% selected the 'maybe' category" (Arbuckle, 2021). According to personal communication with STRIPS project coordinator Omar De Kok-Mercado, at the end of 2020 there were 82 prairie strip sites in Iowa. The project is seeing interest in prairie strip implementation across the country as well, with sites in 13 other states (De Kok-Mercado, 2021). These sites are officially designated by the STRIPS project and/or the USDA's CRP. However, this documentation can't capture possible influence and informal implementation of prairie strips by farmers and landowners who find the paperwork more onerous than the practice of letting a field edge or swampy spot self-seed to prairie.

THE TECHNICAL COMMUNICATION OF THE STRIPS PROJECT

Two stewards of the STRIPS program at Iowa State University, Tim Youngquist and Omar De Kok-Mercado, responded to our questionnaire that explored the project's progress and persuasive strategies the project uses for outreach to landowners and farmers who are curious about installing prairie strips.

Persuasion strategies

Personal attention and compromise between immediate stakeholders are major concerns in the project. The relationship between landlord and tenant is particularly important to the implementation of resilient agriculture and conservation measures, according to both Youngquist and De Kok-Mercado. In Youngquist's experience, the landlord often makes first contact with STRIPS to ask for information and assistance. However, the tenant who rents and does the physical work of farming the land may be more skeptical. For example, Youngquist explains, the tenant farmer is the person who must maneuver large farm equipment through the smaller field spaces created by prairie strips. "Not every landlord/tenant pair is going to be comfortable with the practice at the outset," Youngquist notes. This requires what he calls "prairie diplomacy" to find a compromise solution, such as only buffering the edge of a field rather than integrating strips into the middle of a field. This sort of one-on-one, careful, personal communication is key to the program's success but could also limit widespread implementation of prairie strips. This tension is hinted at in several of the STRIPS project's guiding principles, the last one of which is: "Ensure that the quality of our actions is not compromised with continued growth of the project" (STRIPS, 2021). This example further emphasizes the importance of personal relationships and face-to-face communication in resilient agriculture, particularly with the complications of non-operator landlords.

Appeal of the STRIPS program

The economic argument does work for this conservation practice. In STRIPS, the economic argument relies on the premise that farming "marginal land"—land that is swampy, difficult to access, or has poor-quality soil—loses money. De Kok-Mercado notes, "Most farms will have an area that is low-yielding and sub-profitable for row crops. You can't take out more cash than what you have in the bank and demonstrating that low-yielding acres lose money on inputs can be a good platform for advocating to repurpose those acres." Since prairie strips are a relatively cheap resilient agriculture practice to implement, this argument works. However, invoking and valuing economic reasoning over other types of reasoning is risky in the absence of broad policy support and can undermine other resilient agriculture efforts.

Youngquist and De Kok-Mercado identified a few non-economic appeals of the program that we see reflected in the technical communication documents of the program. One initially unintended but persuasive appeal is the use of a prairie strip as a buffer on which farmers can turn their equipment around at the end of a row. The USDA's CRP rules allow land conserved in the program to still be used for this purpose. However, this benefit is often only possible with the deliberate support of a STRIPS specialist. Youngquist explains, "If the placement of prairie strips does not align with operator equipment and the amount of time needed to complete a farming operation, i.e., planting, is increased, then a landowner may choose not to implement prairie strips. This can be overcome by working closely with the farmer to determine equipment widths, specifically the width of the planter or seeding equipment, and spacing prairie strips accordingly." Nevertheless, this benefit explicitly appears in much of the STRIPS technical communication documents and is even invoked in its logo (fig. 7.6).

The logo features the silhouette of a modern, large-wheeled tractor in green overlaid with images of a meadowlark, grass stem, flower, bee, coneflower, and water droplet. Youngquist and De Kok-Mercado remind us not to overlook the aesthetic and emotional appeal of implementing prairie strips. Youngquist wrote, "I have had several farmers comment on the aesthetic quality of the prairie strips and that the prairie area, filled with blooming flowers and birdsong, has become their favorite part of their farm." STRIPS includes the aesthetic appeal in much of their technical communication; for example, the infographic in figure 7.7 appears on their website.

Figure 7.6. The STRIPS logo.

Figure 7.7. STRIPS in the CRP infographic.

The infographic displays a prairie strip, complete with native flowers, flanked by row-crop fields. The infographic draws attention to a prairie strip's aesthetics through added drawings of birds and native flowers. De Kok-Mercado (2021) links these aesthetics to pride of place. "Natural heritage is also a major factor," he says; "many Iowans are proud to showcase their prairie heritage and incorporating it into the farming landscape is a win-win for Iowa culture."

The third appeal that works for the STRIPS program is, in fact, the environmental argument. Youngquist (2021) notes, "The disproportionate benefits that prairie strips can provide—added water infiltration; soil and nutrient retention; increased pollinator and wildlife habitat; blooming wildflowers; and more are all positives that seem to resonate with farmers and farmland owners." It's important to remember that the relationship between farmland, landowners, farmer-operator-renters, and other stakeholders is a complicated network of at times conflicting beliefs, tropes, and practical needs.

Conclusion

In a recent post to the University of Iowa's Hydroscience and Engineering IIHR blog, "The Land of Milk and Money," water quality research engineer Chris Jones (2020) tells the story of an Iowa grocery store employee's decision to dump 800 gallons of milk down a storm sewer and the resulting public outrage when the local creek "became the Milky Way." Suddenly the creek's pollution, albeit minor, was visible. "But Fourmile [Creek] has never really been more than a conduit for agricultural and urban runoff from northern and eastern Polk County," Jones continues, "and northeast Des Moines proper has suffered devastating floods because of its altered hydrology." Jones implies that sedimentation, nitrate, and dissolved phosphorus, three of the top pollutants in Iowa's waterways, cannot be as easily seen in the water as milk, but they are just as present. Making the invisible visible, and tracing a consequence back to its source, are crucial steps to public awareness for which technical communicators are uniquely trained to contribute. We believe that technical communicators who systematically and continually attend to context can engage and impact positively, even when it seems the odds are stacked against them.

Agricultural decision making often occurs in private, and its consequences are often unseen by both the public and the decision makers. At the same time, those who do have access to agriculture's decision makers can misread the rhetorical situation, discount the communication network's influence, and alienate those whom they most need to reach. We see the opportunity for more and better technical communication as resilience becomes even more important in modern agriculture. We technical writers, teachers of technical communication, science journalists, and conservation advocates have a critical task in front of us: nothing less than turning the ship of Big Ag and Big Petroleum in the direction of policies and practices that bring us cleaner water and healthier soil in the near term. Our declining soil and water quality is an existential threat to humankind, so our work is important.

References

Agriculture's Clean Water Alliance (ACWA). (2016a, June 10). Facebook post. [That's a wrap!]. https://www.facebook.com/AgCleanWaterAlliance/

Agriculture's Clean Water Alliance (ACWA). (2016b, June 10). Facebook post. [Public-private & urban-rural partnership]. https://www.facebook.com/AgCleanWaterAlliance/

Al-Kaisi, M. (2020). Frequent tillage and its impact on soil quality. *Integrated Crop Management Encyclopedia*. Iowa State University Extension and Outreach. https://crops.extension.iastate.edu/encyclopedia/frequent-tillage-and-its-impact-soil-quality

Arbuckle, J. G. (2021). *Iowa farm and rural life poll–2020 summary report*. Sociology Extension, Iowa State University. https://store.extension.iastate.edu/product/16071

Carolan, M. (2008). Democratizing knowledge: Sustainable and conventional agricultural field days as divergent democratic forms. *Science, Technology, and Human Values, 33*(4), 508–528. Retrieved November 23, 2020, from http://www.jstor.org/stable/29734050

Cocklin, C., Mautner, N., & Dibden, J. (2007). Public policy, private landholders: Perspectives on policy mechanisms for sustainable land management. *Journal of Environmental Management, 85*(4), 986–998. doi:10.1016/j.jenvman.2006.11.009

Culver, J. C., & J. Hyde (2000). *American dreamers: A life of Henry A. Wallace*. Norton.

De Kok-Mercado, O. (2021, February 22). Personal email with Lee Tesdell.

Des Moines Water Works. (2020) *History*. http://www.dmww.com/about-us/history/

Eaton, C., Ingelson, A., & Knopff, R. (2007). Property rights regimes to optimize natural resource use: Future CBM development and sustainability. *Natural Resources Journal, 47*(2), 469–496.

Eller, D. (2014). Erosion estimated to cost Iowa $1 billion in yield. *Des Moines Register*. https://www.desmoinesregister.com/story/money/agriculture/2014/05/03/erosion-estimated-cost-iowa-billion-yield/8682651/

Environmental Working Group. (2019). *Iowa's private wells contaminated by nitrate and bacteria*. Iowa Environmental Council. https://www.ewg.org/interactive-maps/2019_iowa_wells/

Haggerty, J. H., Auger, M., & Epstein, K. (2018). Ranching sustainability in the northern Great Plains: An appraisal of local perspectives. *Rangelands, 40*(3), 83–91. doi:10.1016/j.rala.2018.03.005

Harden, Ray. (n.d.). *Des Moines Waterworks lawsuit*. Iowa Citizens for Community Improvement. http://iowacci.org/in-the-news/des-moines-waterworks-lawsuit/

Iowa Partnership for Clean Water. (2015). *What is Bill hiding?* [Video posted on Facebook]. https://www.facebook.com/IowaPartnershipForCleanWater/videos/897764360315892

Jones, Chris. (2020). The land of milk and money. *IIHR blog*. University of Iowa. https://cjones.iihr.uiowa.edu/blog/2020/09/land-milk-and-money

Kamau, Mercy, et al. (2018). Early changes in farmers' adoption and use of an improved maize seed: An assessment of the impact of demos and field days. *African Evaluation Journal, 6*:1. doi:https://doi.org/10.4102/aej.v6i1.278

Love, Orlan. (2010). Serious thought given to whether ag tiling helps or harms. *Gazette.* https://www.thegazette.com/2010/07/18/serious-thought-given-to-whether-ag-tiling-helps-or-harms.

National Science Foundation (NSF). (2020). *Merit review frequently asked questions (FAQs).* https://www.nsf.gov/bfa/dias/policy/merit_review/mrfaqs.jsp#4

Office of the State Archaeologist. (2020). *Little barley.* University of Iowa. https://archaeology.uiowa.edu/little-barley

Pezzullo, P. C., & Hunt, K. P. (2020). Agribusiness futurism and food atmospheres: Reimagining corn, pigs, and transnational negotiations on Khrushchev's 1959 U.S. tour. *Quarterly Journal of Speech, 106*:4, 399–426. DOI:10.1080/00335630.2020.1828605

Pitt, D. (2019). Iowa high court to decide if farm pollution suit continues. *Des Moines Register.* https://www.desmoinesregister.com/story/money/agriculture/2019/11/05/iowa-supreme-court-cci-raccoon-river-lawsuit/4166690002/

Redwave Digital. (2015). *Iowa Partnership for Clean Water TV ad.* https://www.youtube.com/watch?v=MwlHkPDw9s0

Science-Based Trials of Rowcrops Integrated with Prairie Strips (STRIPS). (2021). Iowa State University. https://www.nrem.iastate.edu/research/STRIPS/

Sherman, B. (2015). Tracing the treaties: How they affected American Indians and Iowa. *Iowa History Journal, 7*(4). http://iowahistoryjournal.com/tracing-treaties-affected-american-indians-iowa/

Stowe, B. (n.d.). *External costs and public health threats from nutrient pollution in agriculture watersheds slide set.* Des Moines Water Works. Hosted by Sustainable Food Trust on Slideshare. https://www.slideshare.net/Sustainablefoodtrust/bill-stowe-water

Streit Krug, A., & Tesdell, O. (2020). A social perennial vision: Transdisciplinary inquiry for the future of diverse, perennial grain agriculture. *Plants, People, Planet.* Early View. https://doi.org/10.1002/ppp3.10175

Stroman, D., & Kreuter, U. P. (2016). Landowner satisfaction with the Wetland Reserve program in Texas: A mixed-methods analysis. *Environmental Management, 57*(1), 97–108. doi:10.1007/s00267-015-0596-8

Swinton, S. M., Tanner, S., Barham, B. L., Mooney, D. F., & Skevas, T. (2017). How willing are landowners to supply land for bioenergy crops in the Northern Great Lakes Region? *Global Change Biology: Bioenergy, 9*(2), 414–428. doi:10.1111/gcbb.12336

Tulloch, A. I. T., Tulloch, V. J. D., Evans, M. C., & Mills, M. (2014). The value of using feasibility models in systematic conservation planning to predict landholder management uptake. *Conservation Biology, 28*(6), 1462–1473. doi:10.1111/cobi.12403

Upadhaya, S., Arbuckle, J. G., & Schulte, L. A. (2020). Developing farmer typologies to inform conservation outreach in agricultural landscapes. *Land Use Policy.* https://doi.org/10.1016/j.landusepol.2020.105157.

USDA National Agricultural Statistics Service. (2017). *Census of agriculture*, vol. 1, chap. 2, table 41. Complete data available at www.nass.usda.gov/AgCensus.

Youngquist, T. (2021, February 3). Personal email with Lee Tesdell.

Zhang, W., Plastina, A., & Sawadgo, W. (2018). *Iowa farmland ownership and tenure survey, 1982–2017: A thirty-five year perspective.* Iowa State University Extension and Outreach. FM 1893. https://www.card.iastate.edu/farmland/ownership/FM1893.pdf

Chapter 8

Writing for Clients, Writing for Change

Proposals, Persuasion, and Problem Solving in the Technical Writing Classroom

MONIKA A. SMITH

> In addition to a safe social environment, human well-being requires a safe natural environment. Therefore, computing professionals should promote environmental sustainability both locally and globally.
>
> —ACM Code of Ethics and Professional Conduct (sec. 1.1)

This chapter offers an exciting and, I believe, inspiring story of transformational learning in the technical writing classroom. It maps "what happened" when, in Fall 2016, I had the privilege of introducing a campus-based sustainability project to ENGR 240: Technical Communication, a required course for second-year computer science students at the University of Victoria, a research-intensive university in British Columbia, Canada. How this project came about, what students learned from it, and how writing for environmental action might be conceptualized and practiced in the writing classroom forms the crux of this story.

The science disciplines traditionally viewed stories "as too personal, too incomplete, too anecdotal to qualify as formal research in our field" (Small, 2017, p. 235). Yet "the consequential linking of events and ideas" (Riessman, 2007, p. 5) can offer a meaningful tool for inquiry and scholarship: stories resonate with readers and listeners, gloss data with meaningful social explanations, and offer qualitative insights that supplement and humanize statistical reasoning. As a result, the human sciences (Riessman, 2008; Singleton et al., 2019)—and, in their wake, the physical sciences (Olson, 2015; Suzuki et al., 2018)—increasingly acknowledge storytelling as a form of meaning making that supports "critical reflection, interpersonal dialogue and ultimately, social action" (Singleton et al., p. 22).

My story aspires to contribute a dimension of "social action" to the conversation sparked in this collection. While pedagogically inflected, it focuses primarily on my own observations and reflections, filtering students' experience as learners through that lens. Via that lens, however, I hope to illuminate the value not only of harnessing technical writing projects to environmental action but of localizing such action in a specific, personal space: students' home campus. In this case, students' home campus is a 400-acre "destination campus" nestled in a setting of woodlands, gardens, and meadows within a stone's throw of the Pacific Ocean, located on the traditional lands of the Lekwungen, Songhees, Esquimalt, and WSÁNEĆ peoples, who maintain close connections with the land and the university as students, community members, and knowledge keepers. Indeed, the value of rooting a sustainability-themed client project in a local sense of place—students' campus as home away from home—is, I believe, a key takeaway of this story because it shifts the classroom dynamic from *getting* students to write for environmental justice to *inspiring* them to do so.

In light of Black Lives Matter, technical communication associations have rallied to find ways to support racial, social, and economic justice, issues that need urgent redress. Consider the June 2, 2020 "Call to Action" by Angela Hass (2020), then president of the Association of Teachers of Technical Writing (ATTW), which urged members to use "rhetoric and technical communication skills to redress anti-Blackness in our spheres of influence" (Hass, 2020). The Council for Programs in Technical and Scientific Communication (CPTSC) likewise created a new set of Research Grants to Promote Anti-Racist Programs and Pedagogies (CPTSC, October 15, 2020) aimed at dismantling structures of oppression and discrimination (CPTSC, December 22, 2020). In a similar vein, *IEEE Transactions on Professional Communication* (August 12, 2020) announced a special issue on enacting social justice in technical and professional communication, framing it as a "call to action to respond to TPC's own continued complicity in issues of injustice as well as emerging social justice issues across the globe" (IEEE, 2020).

Environmental degradation, I would argue, is precisely a social justice issue. "Environmental destruction, just like social injustice," states Coates (2007), "is a societal problem stemming directly from the values and beliefs (modernism) that are inherent in the structure of modern society" (p. 213). Contaminated water, polluted air, and toxic waste all burden poor, underresourced communities far more heavily than wealthy, privileged ones, where pollution is not only less conspicuous but also less likely to impact health, livelihoods, and food security (Global Policy Forum, 2020). Human relations inhere in all commodity relations; hence, moves to protect the environment challenge the profit motive that drives the exploitation of people *and* despoils the planet. Advocating for a more sustainable future, in other words, hinges on "developing social structures that support the well-being of all life" (Coates, p. 223), a commitment to equity and justice in its most encompassing sense.

Yet discussions of how technical writing—and, in particular, the *teaching* of technical writing—might be used to spur environmental action as part of a wider "social justice stance" (Jones et al., 2016, p. 211) are few and far between. Jones et al., for example, identify technical writing as a node for interrogating social justice issues focused on exclusion, marginalization, discrimination, and disability. However, although environmental action has received relatively scant attention as a pivot for social justice, groundwork laid by Killingsworth (2005), Linsdell and Anagnos (2011), and others valuably supports Hembrough's (2019) call for an "ecocomposition curriculum" (p. 895) to both enhance professional writing outcomes *and* "interrogate social conflicts located beyond the classroom" (p. 896).

My account therefore joins theirs and others in this volume to energize this conversation. As traced in my story, an ecocomposition writing pedagogy holds much promise for realizing the goals Hembrough (2019) envisions. Working with a client who prompted them to view their campus as *ecosystem* not only raised my students' environmental consciousness; it also encouraged them to become more rhetorically conscious writers, in turn enabling them to revision technical writing as a genre for solving human not just technical problems.

Though situated in reflective storytelling rather than objective study, my account is nonetheless worth sharing, I believe, because it fruitfully intersects with a desire, increasingly voiced, to see more just and humane technical writing scripts both in the professional workplace and the writing classroom. In the context of my classroom, partnering with a campus-based client intent on protecting the local environment offered students a potent means of recalibrating technical writing away from its industrial orientation on "efficiency, expediency, and streamlining processes" (Jones, 2016) toward the "human experience" (p. 344), a move that expanded students' sense of agency not just as technical innovators but as emerging technical writers capable of "enact[ing] social justice" (Colton & Holmes, 2018, p. 4).

The process by which my students came to write for a campus client intent on raising environmental awareness is also worth sharing because it fills a gap in the literature. Rice-Bailey et al. (2020) note that practical accounts of universities that have implemented client-based learning "are uncommon," so my account may also serve as an instructive model for instructors who wish to take a similar approach. Hence, the next stage of this chapter describes the process by which my technical writing students came to take discursive action on behalf of the environment, maps how I negotiated this process, then traces the learning arc students took as *problem solvers* using technical communication genres to shape a more sustainable, earth-centered future.

Technical Writing Requirements and Pedagogies

As computer science majors enrolled in the University of Victoria's Engineering and Computer Science program, my writing students' learning arc begins in ENGR 240: Technical Writing, a one-term "workplace ready" writing course supported both by their faculty and the Canadian Engineering Accreditation Board (CEAB). Qualitative skills that CEAB identifies as graduate attributes, for instance report writing (sec. 3.1.7) and teamwork (sec. 3.1.6), are exactly the kinds of skills ENGR 240 gives computer science majors a chance to practice and develop.

To provide meaningful learning that aligns with communication requirements, ENGR 240's major writing project—a team-written proposal, progress report, and end-of-term feasibility report—originally centered on a fictional case scenario. As Raju and Sankar (1999) note, case scenarios "allow the students to vicariously experience situations in the classroom that they may face in the future" (p. 501), thus serving well as a method of instruction. The context of this and other studies (e.g., Jonassen & Hernandez-Serrano, 2002) was the engineering classroom; however, problem-solving scenarios also work well in first-year composition, professional writing, rhetoric, and business communication courses

(Rosinski & Peeples, 2012; Pennell & Miles, 2009). As Pennell and Miles argue, embedding students' communicative practices—*"rhetoricity, locality,* and *change"* (p. 379)—in realistic case scenarios can spur "deep rhetorical learning" (p. 393).

Nevertheless, a limitation of such scenarios is that they never fully reproduce the real-world conditions and constraints under which professional work, including professional writing, constitutes itself. In Paré's (2002) terms, "no classroom-based simulation of non-academic writing can capture the complexity or intricacy of the original rhetorical context" (p. 59). Further complicating the picture is Yadav et al.'s (2010) finding that a realistic case scenario, while enhancing student *engagement,* did not appear to increase conceptual understanding. This, they speculated, may be due to "the manner in which cases were implemented" (p. 62), implying that best practices for incorporating fictional scenarios in the classroom have yet to fully emerge. Hence, rather than wait for best practices to evolve, it seemed like a good idea to move students beyond the invented scenario.

Stepping toward a Real-World Sustainability Project

Community-engaged learning, which prompts students "to address workplace issues by interacting with real clients in an effort to provide learners with more realistic tasks and environments" (Balzotti & Rawlins, 2016, p. 141), offered a promising way of moving forward. Despite the logistical challenges of setting up "live cases" (Kennedy et al., 2001, p. 147), partnering with real-world clients offers multiple learning benefits. For one, it exposes students to "the expectations of professionals in the workplace [which] can be an education in itself" (p. 147); it centers learning in an "open-ended environment" that moves students toward the "learning-discovery" pole of the learning axis (p. 147); and, in a technical writing context, it can help bolster empathy, preparing STEM students to better meet client, stakeholder, and community needs (Patterson, 2020).

Still, as Balzotti and Rawlins (2016) point out, instructors must "be willing to invest significant time and resources to organize those real-world experiences for students" (p. 141). To ease these demands while still providing students with enriched experiential learning, I turned to our campus community. This proved a wise move, because the University of Victoria not only supports community-engaged learning as a pillar of its Strategic Framework; it also encourages nonacademic units to engage in academic outreach. This meant I would be partnering with a client conveniently located on campus *and* already predisposed to support student learning. Hence, when offered a chance to liaise with Facilities Management (FM) in the summer of 2016, I jumped at the chance. When it comes to finding clients willing to work with students, a campus community offers a remarkably convenient, low-investment opportunity for successful partnerships.

Given the broad mandate of our Facilities Management unit, it was exciting to reflect on the wide array of problem-solving tasks FM might direct students to tackle in their proposal, report, and presentation assignments. Manifold aspects of campus life fall under the FM umbrella: accessibility, buildings and services, capital development, projects, and grounds and maintenance. The unit also works with the university's Office of Planning and Sustainability so is charged with energy and water management, waste reduction, and environmental stewardship of grounds and buildings (University of Victoria, Facilities Management). Consequently, while I felt confident that FM could offer students wide latitude for real-world problem solving, I did not anticipate the singular trajectory it actually took: problem solving for sustainability.

Starting with the energy manager (2016–2017), then the paper recycling manager (2018), and finally the plastics coordinator and supervisor (2019), I partnered my students with a total of three campus clients over a period of four years, each of whom took turns coming into the writing classroom to request students' help in solving environmental issues specific to our campus: how to reduce the university's carbon footprint; how to increase its waste paper diversion rate; and what to do with plastic waste on campus. While the focus on sustainability was unplanned on my part—it emerged, rather, from our clients' own agency (much as it would in the real world)—anyone interested in connecting students with writing opportunities linked to sustainability could do no better than to approach their FM unit as a convenient point of access for tackling local environmental issues. Moreover, while I partnered students with three different FM clients over four years, the fact that they were all affiliated with the same unit meant the consultation process, and the writing deliverables that emerged from that process, remained pretty much constant year to year. This not only valuably streamlined the partnership process; I believe it also attests to the soundness and replicability of that process for anyone interested in adopting a similar approach.

The Consulting Process: Clients and Students

Ensuring a smooth redesign of ENGR 240 was a key goal. I wanted students' transition to experiential learning, plus engaging with a new learning domain—sustainability—to be relatively painless for all involved. To this end, I developed a three-step process of negotiation with my prospective course clients and my students.

STEP 1: NEGOTIATE WITH CLIENTS

The first and most important step was, of course, to liaise with Facilities Management to get the project off the ground. Once FM had decided, as part of their educational outreach, who to "loan out" as a client, the client and I met before the start of term to figure out exactly what students would do to meet course writing outcomes *and* client needs. Each client was actively seeking student input on how to achieve core sustainability goals they were mandated to meet: energy conservation, increased paper recycling, and plastics reclamation. My students' needs, on the other hand, centered on developing proficiency in key technical communication genres—problem definition, proposals, presentations, and reports—all of which were to be produced collaboratively in teams.

To this end, the client and I co-created an RFP focused on a distinct aspect of sustainability for students to respond to. The first RFP, issued by the energy manager in Fall 2016, asked students to propose ideas in response to goals laid out in the university's *Sustainability Action Plan, 2014-2019* (University of Victoria, n.d.):

1. Achieve a total institutionalized greenhouse gas emissions reduction of 30 percent by 2019, relative to 2010 levels.

2. Reduce campus electricity consumption intensity by 8 percent by 2019, relative to 2010 as the baseline year.

3. Reduce campus natural gas consumption intensity by 12 percent by 2019, relative to 2010 as the baseline year.

The second RFP, issued by the paper recycling manager (Fall 2018), likewise focused on the *Action Plan*, but this time asked students to generate ideas on how to increase the university's landfill diversion rate from 71 percent to its 2019 target goal of 75 percent, paying special attention to paper. The following year (2019), in response to China's National Sword Policy banning the import of plastic waste, the plastics recycling supervisor and coordinator issued an RFP asking students to propose "a strategy designed to reduce the amount of plastic used or created on campus or leaving the campus as waste." With their respective RFP in place, students were thus set to solve a local environmental problem for their campus client as part of their major writing project.

Once student and client needs had been aligned through the initiating RFP, the next step was to decide how often the client should meet with students. We anticipated three in-class visits would suffice, though the client also kept avenues of communication open throughout the term. The client's first visit, scheduled early in week two of the course, aimed to familiarize students with two basic elements: the initiating RFP laying out the environmental issue they would tackle and the technical communication genres they would produce (a proposal, progress report, and feasibility report) in the course of developing and formulating their solutions.

We scheduled a follow-up question-and-answer session halfway through the term, just as students were beginning to research the feasibility of the solutions they were proposing. This gave students a chance to request further information, ask questions, and solicit feedback that might prompt them to adapt or modify their ideas as needed. A third and final visit coincided with students' end-of-term progress reports, delivered as a team presentation. For students, this was a highlight of the project. It gave teams a chance to showcase their work directly to the client, report on what they had accomplished, receive feedback, and think about how to shape their upcoming report in keeping with the client's expressed interest in the sustainability solutions they were investigating.

Step 2: Clarify the Nature and Scope of the Project

With audience, topic, and purpose, plus course deliverables, in hand, my next step was to allocate adequate time in class to carefully review the project. This was important, as the project mapped core expectations for what students would do and learn over the next three months. Making time for this step was well spent, because students were tasked with absorbing a lot of new content: a new client-based approach, a new and unfamiliar set of technical communication genres, and a new research topic—resource conservation—that fell outside their intended major.

All this required considerable time to process. Hence, while students were introduced to the client project in week two, they didn't begin actively working on it till week four. Adequate lead-in time, I believe, played an important role in the project's success: it allowed students to process their thinking, reading, and reflection before having to formalize an action plan for creating a cleaner, greener campus. In the end, I found that a well-mapped, well-paced orientation laying out the rhetorical situation helped overcome potential resistance students may have felt on encountering a new, arguably more challenging mode of experiential learning rooted in an unfamiliar area of study: sustainability.

STEP 3: NEGOTIATE WITH STUDENTS

Following the first client project in 2016, student comments on course evaluations expressed some concern that a client project focused on sustainability did not relate to their goals as prospective technical writers whose singular focus would be digital technology. As a result of this feedback, I implemented an extra step: negotiate with students to unfold a metanarrative on the relevance of what they were doing. In the next iteration of the course, therefore, I created space in class to make the case that problem solving for sustainability would benefit students as technical writers, as computer science majors, and as citizens of the world living in "binding and complex relation with others" (Boedy, 2017, p. 117). To make my case persuasive, I relied heavily on appeals to *ethos*, citing professional bodies like CEAB, IEEE, and ACM, which I knew my computer science students respected.

Drawing on these bodies showed students that ecological mindfulness aligns with the values and aspirations of the technical associations they would one day join. This was a winning strategy: as Linsdell and Anagnos point out, "motivating students to practice and perfect their communication skills" (p. 20) is considerably eased when its relevance to future career goals is made clear. To begin, I highlighted CEAB expectations that graduating students should be able to apply "concepts of sustainable design and development and environmental stewardship" (sec. 3.1.9)—terms directly echoing their client's mandate. I also underlined that a commitment to preserving and protecting the environment is enjoined by IEEE, the world's largest technical professional organization, which urges members to "comply with . . . sustainable development practices . . . and to disclose promptly factors that might endanger the public *or the environment*"

(IEEE, 2020, my emphasis). I likewise pointed out that ACM/ IEEE-CS's Joint Task Force on Software Engineering Ethics and Professional Practices directs professional software engineers *and students alike* to abide by eight core principles, the first of which, acting in "the public good" (sec. 1.03), proscribes members from approving software that may "diminish or harm the environment" (ACM/IEEE-CS, 1997). Referencing these bodies helped students to see that far from being irrelevant to their career goals, developing a sustainability mindset could valuably overlap with "learning to think like an engineer" (Buswell et al., 2019, p. 55).

With the orientation and negotiation stages brought to a close, my students were now primed to begin writing for sustainability. The nature and scope of the task they faced, however, was daunting. On the one hand, the Sustainable Campus Index rates the University of Victoria as a "Gold Star performer for sustainability practices" (University of Victoria, About UVic); it has garnered accolades as "one of Canada's greenest employers" (University of Victoria, UVic News, 2020); and it was recently rated by the *Times Higher Education* (2020) University Impact Rankings as number four in the world for action on climate change. Yet the logistics of sustainably managing a growing campus of over 25,000 people studying, living, and working in a total of 45 buildings spread over 400 acres of traditional Indigenous territories are formidable. And it was precisely to remediating these formidable challenges that our campus clients directed my technical writing students' attention.

Focusing on Sustainability: Technical Communication Outcomes

In negotiating with students, I had claimed that writing for sustainability would benefit them not only as technical writers, but as computer science majors and as members of the larger community. Would this be borne out in practice? What were the results of asking my students to take discursive action on behalf of the environment, particularly with respect to a writing course's key measure: communication outcomes? It is to these questions that my story of "what happened" now turns.

OUTCOME 1: AUDIENCE-CENTERED WRITING

One key learning gain of strategizing for sustainability was my students' increased commitment to audience analysis. Attending to "audience, purpose, and occasion," as engineer Mike Alley (2018) puts it, is a vital pre-writing step. According to Ford (2004), "awareness of audience analysis" (p. 303) supports effective writing *and* effective problem solving by prompting students to consider solutions *from a client perspective*. For engineers, this kind of situational awareness is essential. As per the engineering executives cited in Norback et al.'s survey, "more important than what the engineer says is what the audience perceives" (p. 14). Understanding audience goals, values, and needs filters out options not directly relevant to those needs. Consequently, audience analysis not only helps students become more effective writers; it also enhances problem-solving skills, making for more practicable, relevant, client-centered solutions (Patterson, 2020).

Yet students are often resistant to engage with this "zero" stage of the writing process. I speculate this is because in the absence of a "real" audience students view constructing audience profiles as inauthentic. However, presented with a *rhetoricity* and a *locality*, as Pennell and Miles (2009, p. 379) put it, related to their campus home turf, my students seemed keen to engage with audience analysis for what it should ideally be: an eye-opening point of inquiry. Adding the element of "locality" to the rhetorical situation imbued audience analysis with authentic purpose: it meant learning about their campus as *place*—a place, moreover, to which they had affective ties as a home away from home, place of study, and source of friendship, belonging, and community, thus evoking genuine interest.

The interest my students took in learning about their clients as custodians of a place to which they felt personally attached sparked a largely self-directed turn at audience analysis. Meaningful audience analysis takes time and effort, yet as I circulated among students, I observed them put considerable energy into locating, accessing, summarizing, and synthesizing multiple pages and reports housed on the university website relating to their client's stewardship mandate: policy statements on sustainability, principles of environmental stewardship, strategic goals, waste management and recycling initiatives, and budgetary constraints under which power, paper, and plastics on campus were curated.

This is one reason I believe it is so valuable to situate a sustainability project in the context of students' lived experience of campus life. It taps into an affective domain that fuels the labor-intensive process of research that authentic audience analysis demands, a process heavily scheduled STEM students may otherwise resist. Extrapolating further, one could say a project centered on investigating their campus as *ecosystem* nudges students toward a more attenuated responsiveness—or *empathy* in Patterson's (2020) terms—to client and community needs, an aspect that technical writing pedagogies are just beginning to explore and that a place-centered writing project may usefully complement.

Strategizing for a more sustainable campus also drew students to recognize that if they wanted their ideas for waste management to truly make a difference, then those ideas had to resonate not just with their client, but with the larger campus community who would be directly impacted by the changes they envisioned. Community outreach thus became students' next self-initiated step. They surveyed fellow students, contacted support staff, solicited feedback from professors, interviewed campus administrators, and reached out to various suppliers, manufacturers, and distributers in an incipient form of stakeholder consultation. And with stakeholder consultation playing an increasingly prominent role in project planning (Last, 2018), we are doing students a favor if we can prompt them to consider wider stakeholder rights, a step that moves them closer to negotiating the broader terrain of citizen and human rights under which ·the right to a sustainable future falls (Boyd, 2015). By consulting with the campus community on how best to curate waste streams, students were putting into practice the notion that technical writing forms part of an audience-centered community dialogue about what changes are feasible or desirable.

Outcome 2: Enhanced Motivation in the Technical Writing Classroom

Helping students make sense of what they are learning makes for productive outcomes. Learning as "sense-making" may be especially meaningful in a technical writing context where, as Linsdell and Anagnos (2011) note, students often "start the semester with low motivation because they do not feel that the course will benefit them. They do not want another 'English' class" (p. 21). From teaching technical writing courses, which students rarely *elect* to take, I know firsthand how challenging it can be to motivate students not simply to write, but to write well. Moreover, attending to content *as writing* is something my students, like so many STEM majors, often feel poorly equipped to do (Buswell et al., 2019). And yet low writing motivation has consequential impacts beyond the classroom.

Engineering executives report that when hiring they consider an applicant's ability to express ideas clearly, succinctly, and concisely (Norback et al., 2009, p. 14). This is not surprising because putting together a persuasive proposal hinges, in part, on *ethos*—building trust—and, as

Norback et al.'s executives note, trust is built when engineers can explain what they do in accurate, unambiguous, accessible ways. Hence, anything that can motivate engineering students to attend to rhetoric—to the *quality* of their writing—carries weight. Again, this is where writing for sustainability comes into its own. If students are inspired to "write best when interested in the topic" (Linsdell & Anagnos, 2011, p. 22), then the "turn toward place in rhetoric and composition" (Pennell & Miles, 2009, p. 380), especially when that place is students' own campus, shows much promise.

In writing for "actionable knowledge" (Jones et al., 2016, p. 212), in this case greening their campus, one of my students' key takeaways was the importance of *rhetoricity*. In strategizing how to reduce energy and resource intensive behaviors to make their campus part of the solution to climate change, students came to recognize that *rhetoricity*, not just content—*how* they wrote about their ideas, not just the ideas themselves—plays a crucial role in determining whether those ideas get heard. It made sense that to ensure a receptive audience who would approve and implement their ideas, they would need to attend to the textual details framing those ideas: plain language, formatting, headings, topic sentences, signaling. Certainly, a message our campus clients took pains to emphasize was that new and innovative ideas *clearly and concisely expressed* were exactly what they, and budget administrators, were looking for.

Reiterated throughout the term, this message foregrounded rhetoricity as sense making, fueling students' motivation to write well. More than just a hoop to jump through, the labor-intensive process of review, revise, edit, and proofread was adopted as a practical way of turning ideas into action. An informal exit survey by my program colleague, Suzan Last, showed that of 100 students surveyed, 70 percent agreed that working with a campus client enhanced their writing motivation; only 8 percent disagreed, and 22 percent were neutral.

Qualitative feedback yielded complementary results. More than half the survey comments related to students' increased sense of motivation. Enhanced motivation adds value to learning in any setting but is particularly salient in a technical writing context where motivation can be lacking. Qualitative descriptors such as "meaningful," "inspired," and "worthwhile" suggest that writing for a client in the immediate context of their campus inspired students to *want* to write well. The prospect of having "an actual impact on the campus," as one student wrote, was galvanizing: the team wanted to generate a report that was both "accurate" and "feasible." Circulating in class, I noticed students put concerted effort into a process-oriented, *rhetorical* approach to writing, using formative feedback, revision checklists, and self and peer review to achieve their problem-solving goals. Audiences respond best when lines of communication are clear and accessible. Hence, the value of writing projects that motivate students to foreground rhetoricity as a practical way to enact positive change in their small part of the world cannot be overstated.

Outcome 3: Team Cohesion and Communication

From observing my students, I would say that strategizing for a sustainable campus also paved the way for improved team communication. CEAB graduate attributes require engineering students to function competently as "members and leaders of teams" (sec. 3.1.6). ENGR 240 therefore focuses student learning on team building: how to communicate in respectful, inclusive ways using active listening, paraphrasing, constructive feedback, and micro-affirmations. Learning to apply such strategies is vital. As Norback et al. (2009) found, hiring executives look for "active listeners who repeat, clarify and summarize" (p. 14). Getting students to apply these strategies, however, can be challenging, especially in diverse settings.

Pennington (2008) notes that "team actualization depends on goal congruence" (p. 11). Yet individual differences, including diverse temperaments, competencies, interests, and levels of motivation, make goal congruence challenging. Cultural and linguistic diversity adds yet another layer of complexity. At the University of Victoria, international students form around 20 percent of the student body, roughly 4,000 students, with approximately 3,000 of those on study permits from over 100 different countries (University of Victoria, International Perspectives). ENGR 240 draws students from nations all over the world: the United States, China, India, Japan, Germany, Brazil, and Saudi Arabia, to name a few, each of whom bring different needs, politics, perspectives, and languages into the classroom.

From what I saw, however, writing for environmental action seems remarkably well suited for bridging difference. True, environmental degradation and the "environmental burden of disease" (Boyd, 2015, p. e353) are not evenly distributed across the globe. But its ramifications *are* global: a polluted planet spares no one. In the words of poet Marie Howe 2019), "Trashed/oceans don't speak English or Farsi or French."

Irrespective, therefore, of region, nationality, language, culture, ideology, and other identity markers, problem solving for sustainability can offer an end goal—a cleaner, greener campus and, by extension, a cleaner, greener world—that *all* students can relate to and benefit from. Checking in with student teams, I observed key strategies for inclusive communication—active listening, paraphrasing, and shared leadership—being practiced; international and ESL students seemed more willing to take on active roles, speak up, and be heard, while domestic students seemed more willing to listen and less prone to dominating the discussion. Moreover, students from China, India, Germany, and Brazil seemed as invested as their domestic peers in finding ways to remediate waste on *this* campus, on the west coast of Canada, because those same strategies could potentially be adapted to *any* campus. The "importance of place" (Hembrough, 2019, p. 895)—and of preserving place—speaks to a millennial generation with a global outlook.

Writing instructors looking to coalesce student teams around a common goal that resonates with a diverse range of students and that, in turn, encourages inclusive, respectful communication would do well to look to a locally grounded yet globally relevant theme like sustainability. Finding a way to build team cohesion around a shared goal that intersects along multiple differences not only inspires students to apply practical communication strategies; it also allows them to benefit from the enrichment that multicultural perspectives bring to the research table (Freeman & Huang, 2014). And given the "near certainty" (Norback et al., 2009, p. 16) that engineering teams will increasingly operate across multinational lines, incorporating themes that foster communication and collaboration is one of the most important outcomes we can offer our technical writing students.

OUTCOME 4: CROSS-DISCIPLINARY ENRICHMENT

If diverse teams offer expansive learning opportunities, so too does cross-disciplinary exploration. Being "curious about the knowledge of other people" (Leslie, 2014, p. 150) expands the horizon of thought; as Pennington (2008) notes, interdisciplinarity promotes innovative ways of seeing, knowing, and thinking that can serve as a "springboard to creative problem solving." This would seem especially pertinent in a technical writing classroom, where students are learning to write problem-solving documents. In his book on curiosity, Ian Leslie specifically cites digital creators as among those likely to benefit from "the serendipitous collisions" of cross-disciplinary thinking that can erupt in "brilliant ideas" (p. 145).

No wonder CEAB identifies the ability to operate in multidisciplinary settings as a graduate attribute (sec. 3.1.6). For my computer science students, engaging with sustainability frameworks opened new ways of seeing their own computing discipline *and* the role of technical writing.

Figuring out how to reduce the university's greenhouse gas emissions, increase paper recycling, and dispose of plastic waste in environmentally friendly ways meant my students first had to pollinate their existing specialization, computing, with information gleaned from an unfamiliar domain: greenhouse gas emissions, SDGs, waste management chains, conservation values, recycling policies and practices, landfill capacities, and so forth. All this information had to be newly gathered, processed, analyzed, and synthesized from fields outside students' regular area of study, and the effect was to infuse their existing knowledge base with eye-opening insights.

From the energy manager students were surprised to learn that, as of 2016, the university's Data Center—the physical location of which few students were even aware—was not only the second largest consumer of electricity on campus (at around 320,000 kW/h per month) but expected to soon exceed all other facilities (David Adams, personal communication, September 18, 2017). Tracing this to national and global levels, my students further discovered that data centers in the United States consume approximately 2 percent of total US energy and produce roughly 3 percent of its greenhouse gas emissions (Sverdlik, 2016), a figure that globally is estimated at 1 percent of total power consumption (Sverdlik, 2020).

In short, strategizing how to reduce the university's carbon footprint allowed my students to gain deeper awareness of their own field of study, computing. Examining energy- and resource-intensive behaviors on campus through an ecological lens not only brought home the idea of climate change; it also equipped students, as future designers of information technology, with an environmentally conscious lens through which they could envision the emergent Internet of Things as potential nodes of energy conservation: smart grids, energy-smart housing and energy systems, microgeneration, energy positive buildings, and green software (Hilty et al., 2013). One student team, for example, proposed designing a machine learning algorithm for optimizing the Data Center's power usage efficiency (PUE). Problem solving for sustainability can prompt students to connect what they're learning in the technical writing classroom with innovative possibilities in their own discipline, thus heightening its relevance and value.

OUTCOME 5: WRITING FOR CHANGE

When students value what they're learning, they are more likely to engage in learning transfer and apply it in practical ways outside the classroom (Ford, 2004; Kember et al., 2008; Albrecht & Karabenick, 2018). On the one hand, responding to their campus not just as place but as *environment* alerted students to ways in which their campus—and indeed their own major—is implicated in heavy, often wasteful, consumption of energy, paper, and plastic. On the other, it encouraged them to see technical writing as a set of genres in which community health and well-being—people *and* environment—could be envisioned as a core value. Planning for a more sustainable campus gave students permission to recalibrate their sense of technical writing as "responsible social *praxis*" (Surma, 2005, p. 4): a form of caretaking rather than, as traditionally understood, its "usefulness to the employer" (Ornatowski, 1992, p. 91).

Targeting resource waste on campus primarily as problem solvers, students saw little alternative but to put forward some fairly radical proposals. One team, for example, proposed that the university should prioritize sustainable procurement by refusing tenders from suppliers using nonrecyclable packaging (a policy position toward which the university has since moved). Another team explored the viability of repurposing plastic waste for use in 3-D printing. Other teams proposed a variety of Share-a-Mug and/or compostable mug programs, apps for incentivizing students' sustainability practices, sensors to reduce lighting in empty classrooms, a composting center for Food and Housing Services, and a range of awareness-raising programs (e.g., an online sustainability module for all incoming residence students). Seen from this standpoint, students were adopting what technical writing theorists are calling for: communicative practices that "positively impact the mediated experiences of individuals" (Jones, 2016, p. 344), an approach predicated on taking care of the planet itself.

Extrapolating further, harnessing a technical writing project to an ethos of environmental conservation foregrounds a holistic approach to problem solving that could feasibly counter the juggernaut of "economic self-interest" (Ross et al., 2019, p. 2) that exploits resources and people alike. Moreover, looking to long-term wellness for both planet and people doesn't mean replacing students' STEM orientation with a humanities one; the ACM/IEEE-CS Software Engineering Code (1997) anchors its vision of the public good "in the software engineer's *humanity*, in *special care owed to people* affected by the work of software engineers" (ACM/

IEEE-CS, 1997, my emphasis). Though I can only speculate based on what I observed in my classrooms, having students write for a sustainable campus may well nudge them as emerging technical experts to consider an environmentally conscious mindset as crucial for framing meaningful action in a changing physical and cultural landscape.

Lessons Learned: Stories, Observations, and Reflections

From an academic standpoint, my story of "what happened" when campus clients tasked ENGR 240 students with reporting on their ideas for greater efficiencies in curating power, paper, and plastics takes a nontraditional approach to meaning making: it relies on my personal observations and reflections as I circulated among students, talked with them, and watched them explore and draft their assignments. To this extent, it forms part of an emerging conversation, waiting on future studies to provide an empirical perspective on lessons learned. Emerging and exploratory though it is, however, my account supplies a provisional model for experimenting with a similar approach. For one, it suggests that partnering with campus units like Facilities Management can ease the challenge of liaising with external clients, while still providing the enriched learning that writing to solve real-world problems offers. It also suggests that introducing a campus client with a conservation mandate brings home to students the need for responsible waste and resource management in pertinent, pressing, and personal ways.

Enhanced motivation in any classroom, but especially in the technical writing classroom where enthusiasm may be lacking, is desirable because it supercharges learning. Though qualitative in scope, my story supports the idea that anchoring a sustainability project in the lived experience of students' own campus elicits a uniquely affective response. Asking students to deploy audience, genre, and rhetoric as a way to preserve and protect their local environment in "material and perhaps life-changing" ways (Surma, 2005, p. 37) helps boost writing motivation. It may also amplify students' sense of agency as prospective technical writers. As Matthews and Zimmerman (1999) point out, "Students become motivated to accept responsibility for their own education [when] their communication matters—it has direct results in other people's lives" (p. 386).

The story of "what happened" when my students were given the chance to use communication genres to "make a difference" in their small part of the world also aligns with Hembrough's (2019) point that environmental topics that "link students with concerns in their personal lives and majors" can prompt them "to reach out to larger audiences" (p. 909)—exactly the narrative arc my story has traced. Writing for an environmentally sustainable campus moved my students to closely align with client needs, while also paying heed to the larger stakeholder community *and* to each other as members of a team united around a common, globally relevant goal. Ross et al. (2019) claim that when technical communicators adopt a place-based ethos, it not only impels them to deeply consider their immediate and personal surroundings, but it can also prompt them to "find the common ground needed to more richly engage with each other" (p. 7), a claim I believe my students' team interactions well illustrates. Based on my observations, an "ecocomposition model" (Hembrough, 2019, p. 895) can augment team building; ease communication and collaboration across individual, national, and cultural lines; and prompt students to see their campus, yes, as *place* but also as a node in a globally interconnected network of energy and resource flow that personally affects each of them, regardless of region or country of origin. This sense of world-spanning interconnectedness may offer students of diverse backgrounds a way to coalesce around a key idea: that regardless of context and location, the natural world "is to be preserved, defended, and made sustainable" (Johnson-Sheehan & Morgan, 2008, p. 11).

In this way, my story also pivots on "an area of growing interest in engineering education": how to get students to *care* for client and community (Patterson, 2020). Could linking technical communication assignments to a place-based ethos counter the finding, reported in Patterson, of a "trend of disengagement in public welfare . . . discovered in engineering education"? Could strategizing for sustainability prepare students to engage with the kinds of civic-minded thinking Boedy (2017) calls for when he states that phronesis, judicious thinking geared to the public good, should form "part of rhetorical training" (p. 116)? Speculating further, what might be the transfer effect of "place-based writing" (Hembrough, 2019, p. 896)? Until empirical studies can answer such questions, my story intimates that "a place-conscious, ecopoetically informed pedagogy" in the writing classroom might offer "a way into the closed doors of business through the minds and hearts of future workers and managers" (Killingsworth, 2005, p. 370), looking forward to a time when the environment is front and center "in any decision-making process" (Ross et al., 2019, p. 22).

It is precisely Boedy's (2017) "response-able" communicative stance that students, scholars, teachers, and practitioners need to take if twenty-first century calls for greater social justice—and the environment is always a social justice issue—are to be heeded and acted on. And there is no better time to do so than now, when those calls are gaining traction locally and globally. Technical communication has not been immune from reproducing social wrongs. However, repurposing it as a tool for environmental action offers students a meaningful way to not only hone essential communication and problem-solving skills, but to preserve and protect what may be the most fundamental of human rights (Boyd, 2015): the right to a healthy environment.

References

ACM/IEEE-CS. (1997, November). *The joint ACM/IEEE-CS software engineering code of ethics and professional practice* [Webpage]. Retrieved from https://ethics.acm.org/code-of-ethics/software-engineering-code/

Albrecht, J. R., & Karabenick, S. A. (2018). Relevance for learning and motivation in education. *Journal of Experimental Education, 86*(1), 1–10. https://doi.org/10.1080/00220973.2017.1380593.

Alley, M. (2018). *The craft of scientific writing* (fourth edition). Springer-Verlag. https://doi.org/10.1007/978-1-4419-8288-9

Balzotti, J., & Rawlins, J. D. (2016). Client-based pedagogy meets workplace simulation: Developing social processes in the Arisoph case study. *IEEE Transactions on Professional Communication, 59*(2), 13. https://doi.org/10.1109/TPC.2016.2561082

Boedy, M. (2017). From deliberation to responsibility: Ethics, invention, and Bonhoeffer in technical communication. *Technical Communication Quarterly, 26*(2), 116–126. https://doi.org/10.1080/10572252.2017.1287309

Boyd, D. R. (2015). The right to a healthy environment: A prescription for Canada. *Canadian Journal of Public Health = Revue Canadienne de Santé Publique, 106*(6), e353–e354. https://doi.org/10.17269/CJPH.106.5341

Buswell, N. T., Jesiek, B. K., Troy, C. D., Essig, R. R., & Boyd, J. (2019). Engineering instructors on writing: Perceptions, practices, and needs. *IEEE Transactions on Professional Communication, 62*(1), 55–74. https://doi.org/10.1109/TPC.2019.2893392

Coates, J. (2007). From ecology to spirituality and social justice. In J. Coates, J. R. Graham, B. Swartzentruber, & B. Ouellette (Eds.), *Spirituality and social work: Select Canadian readings* (pp. 213–228). Canadian Scholars' Press.

Colton, J. S., & Holmes, S. (2018). A social justice theory of active equality for technical communication. *Journal of Technical Writing and Communication, 48*(1), 4–30. https://doi.org/10.1177/0047281616647803

CPTSC. (2020, October 15). *CPTSC grants for antiracist programs and pedagogies* [Blog]. Retrieved from https://cptsc.org/blog/2020/10/15/deadline-oct-31-cptsc-grants-for-antiracist-programs-and-pedagogies/

CPTSC. (2020, December 22). *Congratulations CPTSC anti-racist programs and pedagogies grant award winners!* [Blog]. Retrieved from https://cptsc.org/blog/2020/12/22/congratulations-cptsc-anti-racist-programs-and-pedagogies-grant-award-winners/

Ford, J. D. (2004). Knowledge transfer across disciplines: Tracking rhetorical strategies from a technical communication classroom to an engineering classroom. *IEEE Transactions on Professional Communication, 47*(4), 301–315. https://doi.org/10.1109/TPC.2004.840486

Freeman, R. B., & Huang, W. (2014). Collaboration: Strength in diversity. *Nature, 513*(7518), 305. https://doi.org/10.1038/513305a

Global Policy Forum. (2020). *Environmental degradation* [Webpage]. Retrieved from https://archive.globalpolicy.org/social-and-economic-policy/the-environment/environmental-degradation.html

Hass, A. (2020, June 2). *ATTW president's call to action* [Blog]. Retrieved from https://attw.org/blog/attw-presidents-call-to-action/

Hembrough, T. (2019). A case study: Focusing on sustainability themes and eco-composition through student blogs in a professional and technical writing course. *International Journal of Instruction, 12*(1), 895–914. https://doi.org/10.29333/iji.2019.12158a

Hilty, L. M., Aebischer, B., Andersson, G., & Lohmann, W. (2013). Preface. *ICT4S 2013: Proceedings of the First International Conference on Information and Communication Technologies for Sustainability, ETH Zurich, February 14–16, 2013* [Application/pdf, Online-Resource]. Retrieved from https://doi.org/10.3929/ETHZ-A-007337628

Howe, M. (2019). Singularity. https://poets.org/poem/singularity

IEEE. (2020, June). IEEE code of ethics (sec 7.8). *IEEE Governance* [Webpage]. Retrieved from https://www.ieee.org/about/corporate/governance/p7-8.html

IEEE. (2020, August 12). *IEEE transactions on professional communication* special issue on enacting social justice in technical and professional communication. *IEEE Professional Communication Society* [Webpage]. Retrieved from https://procomm.ieee.org/ieee-transactions-on-professional-communication-special-issue-on-enacting-social-justice-in-technical-and-professional-communication/

Johnson-Sheehan, R., & Morgan, L. (2008). Conservation writing: An emerging field in technical communication. *Technical Communication Quarterly, 18*(1), 9–27. https://doi.org/10.1080/10572250802437283

Jonassen, D. H., & Hernandez-Serrano, J. (2002). Case-based reasoning and instructional design: Using stories to support problem solving. *Educational Technology Research and Development, 50*(2), 65–77. https://doi.org/10.1007/BF02504994

Jones, N. N. (2016). The technical communicator as advocate: Integrating a social justice approach in technical communication. *Journal of Technical Writing and Communication, 46*(3), 342–361. https://doi.org/10.1177/0047281616639472

Jones, N. N., Moore, K. R., & Walton, R. (2016). Disrupting the past to disrupt the future: An antenarrative of technical communication. *Technical Communication Quarterly, 25*(4), 211–229. https://doi.org/10.1080/10572252.2016.1224655

Kember, D., Ho, A., & Hong, C. (2008). The importance of establishing relevance in motivating student learning. *Active Learning in Higher Education, 9*(3), 249–263. https://doi.org/10.1177/1469787408095849

Kennedy, E. J., Lawton, L., & Walker, E. (2001). The case for using live cases: Shifting the paradigm in marketing education. *Journal of Marketing Education, 2*(23), 145–151. DOI: 10.1177/0273475301232008

Killingsworth, M. J. (2005). From environmental rhetoric to ecocomposition and ecopoetics: Finding a place for professional communication. *Technical Communication Quarterly, 14*(4), 359–373. https://doi.org/10.1207/s1542762 5tcq1404_1

Last, Suzan. *Technical writing essentials.* (2018). [Open source educational resource]. Retrieved from https://pressbooks.bccampus.ca/technicalwriting/

Leslie, I. (2014). *Curious: The desire to know and why your future depends on it.* Basic Books. Retrieved from http://ebookcentral.proquest.com/lib/uvic/detail.action?docID=1681915

Linsdell, J., & Anagnos, T. (2011). Motivating technical writing through study of the environment. *Journal of Professional Issues in Engineering Education and Practice, 137*(1), 20–27. https://doi.org/10.1061/(ASCE)EI.1943-5541.0000032

Matthews, C., & Zimmerman, B. B. (1999). Integrating service learning and technical communication: Benefits and challenges. *Technical Communication Quarterly, 8*(4), 383–404. https://doi.org/10.1080/10572259909364676

Norback, J., Leeds, E., & Forehand, G. (2009). Engineering communication: Executive perspectives on necessary skills for students. *International Journal of Modern Engineering Editors, 10*(1), 11–18. Retrieved from https://ijme.us

Olson, R. (2015). *Houston, we have a narrative: Why science needs story.* University of Chicago Press.

Ornatowski, C. M. (1992). Between efficiency and politics, rhetoric, and ethics in technical writing. *Technical Communication Quarterly, 1*(1), 91–103. https://doi.org/10.1080/10572259209359493

Paré, A. (2002). Keeping writing in its place: A participatory action approach to workplace communication. In B. Mirel & R. Spilka (Eds.), *Reshaping technical communication: New directions and challenges for the 21st century* (pp. 57–80). Lawrence Erlbaum Associates.

Patterson, L. (2020, July). *Engineering students' empathy development through service learning: Qualitative results in a technical communication course.*

[Conference presentation]. IEEE ProCom 2020 International Professional Communication Conference [online], Kennesaw, GA, United States.

Pennell, M., & Miles, L. (2009). "It actually made me think": Problem-based learning in the business communications classroom. *Business Communication Quarterly, 72*(4), 377–394. https://doi.org/10.1177/1080569909349482

Pennington, D. D. (2008). Cross-disciplinary collaboration and learning. *Ecology and Society, 13*(2), 8–21. http://www.jstor.org/stable/26267958

Raju, P. K., & Sankar, C. S. (1999). Teaching real-world issues through case studies. *Journal of Engineering Education, 88*(4), 501–508. https://doi.org/10.1002/j.2168-9830.1999.tb00479.x

Rice-Bailey, T., Leitzke, D., & Hildebrand, T. (2020, July 20–21). *Creating value for STEAM students: Incorporating experiential learning into engineering and technical communication classes through community engagement.* [Conference presentation]. IEEE ProCom 2020: International Professional Communications Conference [online], Kennesaw, GA, United States.

Riessman, C. K. (2008). *Narrative methods for the human sciences* (first edition). Sage.

Rosinski, P., & Peeples, T. (2012). Forging rhetorical subjects: Problem-based learning in the writing classroom. *Composition Studies, 40*(2), 9–33. https://tinyurl.com/yz652z8r

Ross, D. G., B. Oppergaard, & R. Willerton. (2019). Principles of place: Developing a place-based ethic for discussing, debating, and anticipating technical communication concerns. *IEEE Transactions on Professional Communication, 62*(1), 4–23. https://doi.org/IEEE 10.1109/TPC.2018.2867179

Singleton, R., Picado Araúz, M. de la Paz, Trocin, K., & Winskell, K. (2019). Transforming narratives into educational tools: The collaborative development of a transformative learning tool based on Nicaraguan adolescents' creative writing about intimate partner violence. *Global Health Promotion, 26*(1), 15–24. https://doi.org/10.1177/1757975916679553

Small, N. (2017). (Re)Kindle: On the value of storytelling to technical communication. *Journal of Technical Writing and Communication, 47*(2), 234–253. https://doi.org/10.1177/0047281617692069

Surma, A. (2005). *Public and professional writing: Ethics, imagination and rhetoric.* Palgrave Macmillan.

Suzuki, W. A., Feliú-Mójer, M. I., Hasson, U., Yehuda, R., & Zarate, J. M. (2018). Dialogues: The science and power of storytelling. *Journal of Neuroscience, 38*(44), 9468–9470. https://doi.org/10.1523/JNEUROSCI.1942-18.2018

Sverdlik, Y. (2016). *Here's how much energy all US data centers consume* [Webpage]. Retrieved from https://www.datacenterknowledge.com/archives/2016/06/27/heres-how-much-energy-all-us-data-centers-consume

Sverdlik, Y. (2020, February 27). *Study: Data centers responsible for 1 percent of all electricity consumed worldwide* [Webpage]. Retrieved from https://

www.datacenterknowledge.com/energy/study-data-centers-responsible-1-percent-all-electricity-consumed-worldwide

University of Victoria, About UVic. (2020). *Organization and governance* [Webpage]. Retrieved from https://www.uvic.ca/faculty-staff/organization-governance/about-uvic/index.php

University of Victoria, Facilities Management. (n.d.). https://www.uvic.ca/facilities.

University of Victoria, International Perspectives. (n.d.). https://www.uvic.ca/about-uvic/about-the-university/international-perspectives/index.php

University of Victoria. (n.d.). *Sustainability action plan: Campus operations 2014–2019*. https://www.uvic.ca/sustainability/assets/docs/policy/action-plan-2014.pdf

University of Victoria, UVic News. (2020, July 6). *UVic one of Canada's greenest employers* [Webpage]. Retrieved from https://www.uvic.ca/news/topics/2020+greenest-employers+news

Yadav, A., Shaver, G. M., & Meckl, P. (2010). Lessons learned: Implementing the case teaching method in a mechanical engineering course. *Journal of Engineering Education, 99*(1), 55–69. https://doi.org/10.1002/j.2168-9830.2010.tb01042.x

Chapter 9

Health in the Shale Fields

Technical Communication and Environmental Health Risks

Barbara George

The relationships between environmental justice and human health revealed through technical communication (TC) about environmental risks are complex, and environmental health concerns are not fully captured by existing technical communication patterns. This is particularly clear in emerging industries such has high volume hydraulic fracturing (HVHF). Because unclear definitions of environmental risk impact long-term decision making about an uncertain industry, public participants affected by the shale gas and oil industry cannot meaningfully deliberate risks about HVHF. HVHF is often framed as a "safe" industry managed by environmental regulatory institutions. But disparities exist between technical definitions and policies that frame environmental risk, place, and human health. There are many interrelated processes of fracking: drilling (which includes physical and chemical components), storage, and transportation. As a result, there are risks associated with various processes: earthquakes, chemicals spills, water contamination, leaking methane, and air quality are just a few. Varied technical definitions of HVHF risk and varied adoptions of a precautionary ideology result in HVHF policies that have uncertain and uneven human and environmental health impacts. It is increasingly clear that effective technical communication for environmental action should challenge the ways environmental risk has been conceptualized and enacted since industry and regulatory agencies often make minimal attempts to widen notions of long-term risk. It is also clear that enacting new policies within existing ideological frames is difficult.

To highlight new ways that technical communication might enact practices more conducive to environmental and human health, this chapter explores and compares the communication of two environmental regulatory agencies—the Pennsylvania Department of Environmental Protection and the Pennsylvania Department of Health—with communication of the Southwest Pennsylvania Environmental Health Project. To accomplish the comparison, I apply Fahnestock and Secor's (1985) concept of stasis theory—which details the connection between definition, value, and policy of given phenomena—to the three organizations to reveal ways that ideologies embedded within technical communication must be considered more broadly for more effective environmental action. I consider how TC can integrate intersectional environmental justice and precautionary ideologies to address stakeholder health risk concerns more fully. In doing so, I ask, "What happens when places do not 'define' industrial practices as a risk through existing technical language?" I look at the Marcellus Shale fields of southwestern Pennsylvania that have been tapped for gas production to ground this analysis.

While, ideally, public participants should be able to access technical communication about human health as impacted by HVHF, a lack of a coordinated effort between environmental, health, and industry institutions in Pennsylvania hampered early attempts to understand and communicate human health risks. Instead, regulatory institutions focused on communicating and enacting economic development, which led to a widespread network of HVHF becoming the status quo in southwestern Pennsylvania in the span of a few short years despite many stakeholders questioning the risks associated with this industry. According to Grabill and Simmons's (1998) model of social constructions of risk, ethical and communicative efforts that result from linear communication models create definitions of risk that often exclude important stakeholders. Similarly, Erin Frost (2015) discusses technical communication scholarship that supports the understanding that risk is socially situated. Because risk is socially situated and not "neutral," a tension exists between decontextualized texts and processes produced by environmental regulatory organizations and those texts participants use when attempting to understand environmental risks that impact their health.

In the first part of this chapter, I outline how environmental health regulatory organizations like the Pennsylvania Department of Environmental Protection (PDEP) and the Pennsylvania Department of Health (PDH) have historically framed HVHF health issues as "status quo" through language that lacks environmental justice and that omits intersectional and precautionary language. By asking a series of questions based in stasis—*fact* (Do environmental health risks exist?); *definition* (What are the environmental health risks?); *value* (Should we act on environmental health risks?); *policy* (How should we act regarding environmental health risks?)—one can understand how precautionary arguments were driven from technical communication about HVHF.

As a positive alternative for technical communication action, I then apply stasis theory to technical documents generated by a local community health and environmental group, the Southwest Pennsylvania Environmental Health Project, to show examples of coordinated, intersectional public justice efforts that reveal complicated interactions between the environment and human health. The organization's site encourages more inclusive, intersectional, and environmental justice language in public deliberation efforts. This analysis of a community health nonprofit reveals a model for technical communication that could be adopted by environmental regulatory and health advisory agencies so that they can be more responsive to shifting knowledge about environmental risks of any industry (though it should be noted that the Southwest Pennsylvania Environmental Health Project's attempts for more just policies are situated in a state that routinely does not consider precautionary action in terms of HVHF).

Literature Review

Current models of environmental risk deliberation have been designed to keep certain stakeholders out of creating definitions about environmental and human health risk. Voicing "risk" about HVHF is difficult. In "Communicating Activist Roles and Tools in Complex Energy Deliberation," George (2019a) discusses the ways that the Energy Policy Act of 2005 deregulated the emerging oil and gas activities from federal oversight laws and placed the burden of proof directly on public participants to show "extraordinary circumstances" to report on or represent HVHF risks (Brady, 2011)

By contrast to a linear model of risk communication, environmental risk communication scholars offer ways to address the complexities of risk and human health through a wider notion of technical communication. Simmons (2007) notes the opportunity for a more inclusive position of stakeholders "by creating the institutional space within which risk can be collectively constructed and more effectively communicated" (p. 437). Similarly, expanded notions of science, including long-term human health, can better assist technical communicators in assessing complex environmental risk. Scholars call for a more complex notion of post-positivist TC that includes extended peer communities and concepts like "sustainability" that contain many stakeholders to address complex, costly, and potentially lethal uncertainties (Funtowicz & Ravetz, 1993; Goggin, 2009; McGreavy et al., 2012; Herndl & Cutlip, 2013; Druschke, 2014). Creating praxis for environmental deliberation means seeking wider possibilities within technical communication about complex environmental risk, including environmental and human health. Technical communication practices must include diverse stakeholders as they discuss complicated industries like HVHF. This means including ideologies of environmental justice, intersectionality, and precaution.

Finally, environmental justice concepts are particularly helpful in expanding ideas about technical communication. In discussing long-term, situated risks, new models of communication acknowledge "the link between human bodies as situated in geographical spaces that are ignored in institutional risk reporting policies and practices" (George, 2019a, p. 42). Applying environmental justice ideologies to analysis of agencies enables a critique of patterns that powerful interests use to exploit marginalized locales for resources, a process that often results in long-term degradation of local land, water, and human bodies. These industries and the agencies that support them often have only superficial exposure to risks but profit greatly from access to resources. Both profit and environmental risks are unequally distributed as local sites deal with long-term, costly, and complex effects of pollution. Environmental justice research explores expanded notions of how citizens might contribute to environmental risk reporting by understanding justice in several arenas: distributive economics, toxins and pollution distribution in space, and toxin and pollution risks through time—all of which have human health consequences (Holifield et al., 2010; Nixon, 2011; Walker, 2010). As George (2019a) argues, these intersections, or "layers" of risk, are often not part of environmental regulatory deliberation about new industries.

Contextualizing Place: Southwestern Pennsylvania—
Past and Current Environmental Policy

As the analysis that follows shows, a disparity of access to technical environmental information exists for different stakeholders deliberating about environmental concerns within specific locations. More ethical technical communication is not a matter of simply changing technical language on a website; it suggests an understanding of the social situations from which deliberative practices emerge in particular places. For example, my past research details how the lack of public participation about the HVHF industry has had disproportionate impacts on residents who live near HVHF industry sites in areas of Appalachia, including southwestern Pennsylvania. These places have histories of extractive economies and poverty and currently lack agency to speak about environmental justice and intersectional concerns (George, 2019b). The disparities are highlighted when considering that an adjacent state, New York, banned HVHF drilling based on a public health argument presented by the New York Department of Health. How could definitions of HVHF risk and attendant policy vary across state lines and across two state health departments (George, 2019b)?

While New York banned HVHF based on a human health argument, in Pennsylvania a state gag order enacted in 2012, Act 13, initially did not allow for disclosure of HVHF chemicals to doctors in cases of suspected exposure. While several provisions of Act 13 were struck down by a 2016 Pennsylvania Supreme Court ruling, years of positioning economic interests above human health effects culminated in a green light for industry to practice HVHF with little concern for precautionary practice (Phillips, 2016). The result is an instance of "slow violence" (Nixon, 2011) that manifests in human bodies, often in low-income, rural areas with a history of exploitive carbon extraction practices. These places and bodies serve as collateral damage to the advancement of the HVHF industry. While the current (in 2021) governor of Pennsylvania, Tom Wolfe, recently called for public health studies related to HVHF, this is only after an outcry based on an increase of widely publicized health studies that suggest HVHF causes a litany of health issues such as childhood cancers and low fertility rates, to name just two. Meanwhile, several chemicals used in HVHF activity are still protected from public knowledge, and powerful interests maintain the status quo of drilling: "The Marcellus Shale Coalition, a gas industry group, maintains there is no credible link between fracking and childhood cancers" (Tanenbaum, 2019).

The historical lack of environmental concern in Pennsylvania underscores the ways in which Pennsylvania's natural identity is often seen, rather narrowly, in terms of carbon extraction. Much of the "story" of Pennsylvania's environmental history relies on carbon extraction and the state and federal policies that exist in response to the pollution of that extraction. Industrial pollution through coal mining, for example, was simply part of the makeup of Pennsylvania cities as extractive industries progressed, and concerted policies to control it were not often put into place until after large-scale public health damage (Hardy, 2012; Black & Ladson, 2012).

Pennsylvania continues to grapple with issues related to abandoned mines throughout the state, including water quality issues, acid drainage from mines, and the struggle for larger economic redevelopment. The federal Surface Mine Control and Reclamation Act (SMCRA) was passed in 1977 to address mine abandonment. However, the implementation of such laws is often ambiguous, and thousands of acres of abandoned mine land in Pennsylvania remain, much of which is now used for HVHF activity, resulting in layered risks.

The lack of "precautionary" language within deliberations for supposedly "clean" innovations like HVHF means toxic, carbon-based industries will continue to impact human health. Pennsylvania's environmental identity is deeply entwined with its history of energy extraction, but the geographic distribution of risk is clustered in the northeastern and southwestern parts of the state, partly dictated by the way in which the possible recoverable shale gas occurs in those parts of the state.

The marginalization of Pennsylvania's landscape and the bodies that live on that land have given rise to some local advocacy groups like the Southwest Pennsylvania Environmental Health Project, which defines itself as follows: "The Environmental Health Project (EHP) is a nonprofit public health organization that assists and supports residents of southwestern Pennsylvania and beyond who believe their health has been, or could be, impacted by unconventional oil and gas development (UOGD, or fracking)" (Southwest Environmental Health Project, 2021). The Southwest Pennsylvania EHP seeks a concerted public health response because "the lack of objective, reliable data on the health effects of UOGD activities raises many questions about the origins of residents' health problems and the scope of public health risks in communities" (Southwest Environmental Health Project, 2021). It is here that technical communication delves into complexities between human bodies and natural gas extraction, both of which are environmental justice and precautionary concerns.

Methods

Analyzing the ways in which the "is" is defined by various actors in HVHF risk leads to understanding the ways socially situated assumptions of values often appear in discussions of risk, and by extension, the "ought," of HVHF policies and practice informed by these definitions (Schiappa, 1998). It is the discrepancy between the "is" and the "ought" that leads to different definitions of risks and different policies and practices that can be applied across language used by various stakeholders.

The way the "is" sets the groundwork for the "ought" underscores Fahnestock and Secor's (1985) suggestion to apply stasis theory as a deliberative tool to examine the way technical communication frames risk and either action or inaction: ideologies in language frame particular policies. For example, Dryzek (2005) suggests that environmental discourses are ideological in that different "discourse rests on assumptions, judgments, and contentions that provide the basic terms for analysis and debate" (9). These discourses and polices about what action to take, then, link to whether a phenomenon is considered a risk or not. Many participants in Pennsylvania are frustrated not only by varied definitions of risk in HVHF but also by questions supported by industry, for example, whether HVHF entails risk at all. Citizens also find frustrating the process by which the HVHF industry might make a claim about risk as a fact even *before* definitional claims of risk are publicly debated. This distinction is important as public participants in the state routinely note the lack of public space to deliberate the existence of risk that suggests stasis theory as a good method. By asking a series of questions based in stasis—*fact* (Do environmental health risks exist?); *definition* (What are the environmental health risks?); *value* (Should we act on environmental health risks?); and *policy* (How should we act regarding environmental health risks?)—stasis theory reveals value orientations among varied stakeholders and the implications of those orientations on differing HVHF policies.

I apply stasis theory to three websites to analyze how environmental risk is represented through technical language and how values inform eventual inaction or action. Two websites are managed by governmental environmental regulatory and health advisory organizations that uphold the status quo within the energy industry. The third is a community health resource that challenges assumptions about human health risk. The first agency I explore, the Pennsylvania Department of Environmental Protection, is the statewide regulatory entity that oversees environmental stewardship in Pennsylvania. Its mission states: "The Department of Environmental Protection's mission is to protect Pennsylvania's air, land, and water from pollution and to provide for the health and safety of its citizens through a cleaner environment" (PA DEP, 2021). While it is noteworthy that protecting human health is a goal of this group, finding specifics about definitions of public health risks as related to HVHF are challenging to find on this site. The next site I explore, the Pennsylvania Department of Health, is the health regulatory body of the state and its mission is "to promote healthy behaviors, prevent injury and disease, and to assure the safe delivery of quality health care for all people in Pennsylvania" (PA DOH, 2021). However, the way this agency continues to distance itself from HVHF risk concerns despite emerging studies is problematic. Both regulatory sites routinely minimize risk through technical language found on their websites. The third website I analyze is the Southwest Pennsylvania Environmental Health Project, which explicitly links human health concerns directly to the HVHF industry, offering a broader construct of risk.

SAMPLE ANALYTICAL PATTERNS

In analyzing language on regulatory and nonprofit sites within the stasis levels, I call attention to the *is/ought construction*. The "is" distinction helps to focus on questions of fact and definition (*Do* health risks exist and what are they?). The "ought" distinction helps to illuminate questions of value and jurisdiction (If a health risk has been defined, what *ought* to be done about it?). Of particular interest is the notion of values informing policy. Various discourses, which Dryzek refers to as "storylines" (2005, p. 17), rest on assumptions that can be applied to stasis and that enable shared understandings, though discourse patterns may vary by place. For the "values" (ought) level of stasis, two distinct ideologies emerged: *administrative rationalism*, which supports status quo HVHF industry extraction practices and *green politics*, which contrasts administrative rationalism by offering complex considerations about risk in HVHF practices.

I apply an example of the stasis framework in table 9.1 to reveal how New York and Pennsylvania represented contested definitions of public health risks of HVHF. The sample analysis reveals varied storylines with

Table 9.1. Stasis and Health Risks: Two States' Health Departments

	PA pre-2016	PA post-2016	NY
Fact (Is) Are there health risks related to HVHF?	PA Act 13 acknowledges there are some health risks	PA overturns Act 13, acknowledges that there are more health risks than previously supposed	Yes
Definition (Is) What are the health risks?	Unclear	Unclear	Unknown
Value (Ought) Should there be action based on health risks?	Unclear	Calls for more health and environmental studies	Ban HVHV drilling; calls for more health and environmental studies
Jurisdiction (Ought) What is the action suggested based on what authority?	HVHF industry continues, relying on DEP regulation; medical staff bound to gag order to protect "trade secrets" of industry; no state health department representation Storyline: Administrative Rationalism	HVHF industry continues, relying on more pointed DEP regulation; stops "gag" rule, but medical regulatory agencies are still not part of meaningful discussions about risk Storyline: Administrative Rationalism	NY DEC bans HVHF in state of New York largely due to health concerns (references to the New York Department of Health) Storyline: Green Politics

different value orientations. An important note about the sample analysis in table 9.1: many of Pennsylvania's Act 13 rules protected industry in terms of eminent domain through the doctor "gag rule," which did not allow medical professionals to access information about chemicals used in fracking in investigating health issues, and through the "exclusion of private wells from the notification of hazardous spills." These protections were overturned in 2016, so table 9.1 offers a pre- and post-2016 stasis analysis of Pennsylvania's approach compared to New York's fracking ban to demonstrate how different places can have widely different environmental policies.

Table 9.1 reveals how definitional concerns about health risks vary, as do the authorities that guide human health policy as related to environmental health and action. In comparing health department statements about risk from Pennsylvania and New York, one can see that each state defines risk differently, an "is" concern. Participants in Pennsylvania have long reported that there is a lack of representation or deliberative space to explore or report bodily harms caused by HVHF. New York State, by contrast, based the HVHF drilling ban on definitional concerns of human health risk as determined by the New York Department of Health. In New York's case, a human health department was part of environmental deliberation that led to specific action: banning HVHF. The distinctions among states that do or do not allow for HVHF drilling activity based on definitions of human health risk are crucial to understanding the socially situated nature of risk. The acknowledgment in New York that public health "is" a risk concern, and that public health is valued as a risk indicator, guides the HVHF ban—an "ought" or actionable concern. This brief, sample stasis analysis—a method more fully employed to examine the three websites addressed in this chapter—shows how stasis theory can reveal definitions of risk and consequent policy actions that either ignore intersectional environmental justice and precautionary language or support it.

Findings and Analysis

The Pennsylvania Department of Environmental Protection and the Pennsylvania Department of Health have communicated largely with an "administrative rationalism" ideological approach. They have done so by the ways that they define the "is" associated with HVHF—that is, that HVHF is *not* a human health concern but *is* good for the economy. Consequently, decisions were made to pursue or continue HVHF activity before public discussions of intersectional environmental justice and precaution could be articulated. Not considering human health risks before the industry commenced nor clearly defining what those risks might be in an unknown industry resulted in a push toward HVHF activity that was often unregulated. Though there was some success in overturning parts of ACT 13, which initially protected the industry to the detriment of human health, the burden of proof to show risk still falls on concerned publics rather than on governmental regulatory institutions.

RESTRICTIVE DISCOURSE: PENNSYLVANIA DEPARTMENT OF ENVIRONMENTAL PROTECTION

Although its name implies protection, the Pennsylvania Department of Environmental Protection (PA DEP) often stymies discussions about environmental risk. The Pennsylvania Department of Environmental Protection is a statewide entity that oversees environmental stewardship in Pennsylvania with a mission to "protect Pennsylvania's air, land and water from pollution and to provide for the health and safety of its citizens through a cleaner environment" (PA DEP, 2021). Technical language constructs about HVFF risk, however, repeatedly compromise that mission.

Stasis Level of Fact

Prior to 2016, there were few acknowledgments about HVHF risk on the website. Instead, much of the information on the site pointed to leasing and regulation of assumed "safe" HVHF activity and navigating the site was difficult for public participant users (George, 2019b). Hence, "action" was business as usual. At that time, health studies shared on the site only minimized potential health risks. For example, a PDF link from 2012 report entitled "Long-Term Ambient Air Monitoring Project Near Permanent Marcellus Shale Gas Facilities" outlines some studies about HFHV activity. On the stasis level of *fact* (Are there health risks related to HVHF?) the report concludes that in the short term, little evidence links HVHF to health risks (fig. 9.1).

While the DEP does acknowledge that sampling was limited, there were no calls at that time for more studies before HVHF activity resumed. The study was finalized in 2018. Using data from previous years, the

- Results of the limited ambient air sampling initiative conducted in the southwest region did not identify concentrations of any compound that would likely trigger air-related health issues associated with Marcellus Shale drilling activities.
- The PA DEP was unable to determine whether the potential cumulative emissions of criteria pollutants from natural gas exploration activities will result in violations of the health and welfare-based federal standards. (DEP, 2012, Long Term).

Figure 9.1. PA Department of Environmental Protection, fact.

"PA DEP Long-Term Marcellus Ambient Air Monitoring Project" also appears on the DEP site. However, this report is not linked to general information about Marcellus Shale. It is, instead, found under a webpage entitled "Monitoring Toxic Pollutants," which a public user might not find since the topics might not be readily connected by nonexperts. Also, only one definition of toxicity is explored in the report—possible air pollut-ants—when a wide variety of possible pollutants were noted by various stakeholders, including water and soil issues (fig. 9.2).

Stasis Levels of Definition, Value, and Jurisdiction

Because the few studies that were enacted did not show "impacts" (What are the health risks as related to HVHF?), the DEP's only recommendation is for jurisdiction (What action is suggested based on what authority?), to continue with HVHF activity while monitoring. It is here that the studies' limitations are clear because the DEP recognizes that the research did not actually apply to those who are most impacted by HVHF (fig. 9.3). It is extraordinary, perhaps, that a study would not monitor the human health of those likely to be most impacted by HVHF from the outset.

Based on these passages, it appears that the value of these decisions (Should there be action based on health risks?) is based in administrative rationalism. This "is" assertion, that there are no health risks, informs the

Observed ambient concentrations of screened toxic and hazardous air pollutants at all five project sites were generally comparable to a low-impact background monitoring site located near Arendtsville, Pennsylvania, in Adams County. This rural, largely agricultural comparison site is unimpacted by any oil and gas exploration, drilling, extraction or processing infrastructure (DEP, 2018, Long-term, p. 117).

Figure 9.2. PA Department of Environmental Protection, fact.

...this project did not examine potential acute or chronic impacts to individuals working in, adjacent to, or in the immediate vicinity of natural gas extraction, gathering and/or processing facilities. The Department should use this and future monitoring data to support efforts by the public health/industrial hygiene community to assess these potential localized risks.

Figure 9.3. PA Department of Environmental Protection, jurisdiction.

"ought"—continued HVHF industrial practice with some monitoring, but no suggestions about banning or modifying the practice.

After many parts of Act 13 were struck down by the Pennsylvania Supreme Court in 2016, the DEP website was redesigned and did attempt to add some participatory links. Currently (in 2021), the site contains a prominent "Submit an Environmental Complaint" button at the top of the DEP page that did not exist earlier in the HVHF process. This links to an online environmental complaint reporting form. While this does offer a communicative link for concerned citizens, specific types of environmental concerns, such as human health concerns possibly related to HVHF, are not suggested, even within environmental justice sections of the webpage.

Throughout this site, the assumption is that permits for HVHF *ought* to be given—and that public participation is included as part of an inevitable status quo of HVHF activity. Similarly, no links appear on this site to describe the potential human health concerns of HVHF, nor do links appear to community resources that might alleviate human health issues related to HVHF. In other words, human health issues related to HVHF are neither recognized nor defined on these pages, so the "ought" distinction (If a health risk has been defined, what ought to be done about it?) is moot. Discussion about possible risk seems to be in name only, not in actual practice, and the underlying assumption is that HVHF will continue with little discussion about the emerging research human health effects of HVHF.

RESTRICTIVE DISCOURSE: PENNSYLVANIA DEPARTMENT OF HEALTH

The Pennsylvania Department of Health (PA DOH) also limits discussion about the risks of HVHF. The mission of the Pennsylvania Department of Health is to "to promote healthy behaviors, prevent injury and disease, and to assure the safe delivery of quality health care for all people in Pennsylvania" (PA DOH, 2021). Prior to 2016, there was little discussion on the PA DOH website about human health risks related to Marcellus Shale, in part because "trade secrets" of chemicals used in HVHF super-seded the possibility of medical inquiries by doctors due to the so-called "gag" rule. In fact, reports exist of PA DOH employees instructed to not engage in public inquiries about possible connections between HVHF and public health concerns (Colaneri, 2014). As studies of short- and long-term health impacts related to HVHF began to surface, the Pennsylvania Department of Health did begin to address possible health issues on their website. However, the burden of proof to demonstrate human health risks of HVHF fell to concerned citizens, not to industry or government (as many physicians who sued to overturn the gag order testified).

Stasis Level of Fact

Despite past distancing from HVHF health concerns by the PA DOH, there is now (in 2021) a section on their website titled "Oil and Natural Gas Production Health Concerns." Here, PA DOH acknowledges the pos-sibility of health concerns related to HVHF, an "is" concern, by including a health registry that allows individuals to report health concerns related to unconventional oil and natural gas development (UONGD) through the Division of Environmental Health Epidemiology. To answer stasis con-cerns at this level (Are there health risks related to HVHF?), the answer is yes (fig. 9.4). In fact, the PA Department of Health links studies to the webpage so concerned citizens can access them.

Stasis Level of Definition

While health risks are acknowledged, the next level of stasis, definitions (What are the health risks?), are less clear. This distinction is important in understanding how this agency positions itself in the HVHF deliberation. For example, the language points to "limited evidence" of UONGD and infant health and suggests that initial research shows no link between childhood cancers and UONGD, although the site does acknowledge the need for more long-term studies (fig. 9.5). Here, studies are situated in a variety of health risk factors linked to HVHF, but seemingly not enough conclusive evidence exists, for example, to ban the activity.

Most epidemiologic research to this point has compared the health outcomes of those living varying distances from unconventional well sites as a substitute for exposure to UONGD. There have been very few studies that have measured exposure directly [link 1]. Overall, epidemiologic work has found some limited evidence of relationships between living near UONGD and poor infant health [links 2-6] and worsening respiratory symptoms [links 7-8] (PA DOH, 2021, Oil and Natural Gas).

Figure 9.4. PA Department of Health, fact.

Although an early county-level study found no association between UONGD and children's cancer incidence in Pennsylvania [link 9] the effect of UONGD on cancer and cancer-related deaths is still unknown. In general, the latency period (the time lag between when exposure occurs and diagnosis) for cancer is at least a few years. Over the next several years, researchers will be able to more accurately study exposure to UONGD and cancer incidence as the latency periods are reached (PA DOH, 2021, Oil and Natural Gas).

Figure 9.5. PA Department of Health, definition.

Stasis Levels of Value and Jurisdiction

The DOH website includes links to resources, including an updated report published by Concerned Health Professionals of New York, entitled *Compendium of Scientific, Medical, and Media Findings Demonstrating Risks and Harms of Fracking (Unconventional Gas and Oil Extraction)* (2020). If a user clicks on this resource, the text points to what ought to be done (Should something be done? What should be done?): "Our examination uncovered no evidence that fracking can be practiced in a manner that does not threaten human health directly and without imperiling climate stability upon which public health depends." In this New York Department of Health statement, found as a "resources" document on the PA

Department of Health website, the suggestion, from a neighboring state, is to halt the practice of fracking at the "ought" level of stasis.

However, despite the link in this statement appearing on the PA Department of Health website, this "green politics" attitude has not made a meaningful impact on HVHF activity in Pennsylvania. Instead, the PA DOH presents many other fact sheets that suggest monitoring for continued HVHF activity will continue through more studies, a position that shows an administrative rationalism ideology. For example, while there is an acknowledgment in health studies that Ewing's tumors could be a public health issue related to HVHF, the PA DOH website states these cases are not "statically significant," indicating that that these findings will not inform upper levels of stasis ("ought") to halt HVHF. This approach is once again couched in the administrative rationalism ideology where experts monitor risks that are "manageable."

While it is noteworthy that this website suggests intersecting health risks are possible, there is neither discussion of precautionary approaches to HVHF deliberation nor any approach to addressing possible risks with HVHF. Equally important, the Pennsylvania Department of Health defers to the DEP for environmental risk health definitions based on limited definitions. For example, a statement on the general PA DOH page that asserts that another agency, the DEP, holds primacy in regulation of HVHF. Unlike New York, whose health experts' research about HVHF resulted in an HVHF ban (the "ought" in the upper level of stasis), the current PA Department of Health avoids such bold statements (fig. 9.6) by ceding judgment to the DEP.

Instead of acting as an agency that can contribute to deliberations and understandings of public health, both agencies' approach to HVHF risk means many concerned Pennsylvania citizens have felt as though HVHF health issues have not been properly addressed because no public deliberation platform exists that usefully combines experts from many agencies.

The Department of Health, however, is an advisory agency only. The Department of Environmental Protection is the state agency that regulates oil and gas activities in the state and collects environmental samples. (PA DOH, 2021, Oil and Gas).

Figure 9.6. PA Department of Health, jurisdiction.

Encouraging Discourse: Southwest Pennsylvania Environmental Health Project

By comparison to the public agencies previously described, the Southwest Pennsylvania Environmental Health Project positions itself differently by aligning itself more closely with intersectional environmental justice approaches.[1] This position offers a possibility for ethically depicting risk through effective technical communication. This nonprofit's mission reads: "The Environmental Health Project (EHP) is a nonprofit public health organization that assists and supports residents of Southwest Pennsylvania and beyond who believe their health has been, or could be, impacted by unconventional oil and gas development (UOGD, or fracking)" (Southwest Pennsylvania Environmental Health Project, 2021). More so than any of the previous sites, this group offers an unequivocal description of health risk related to HVHF.

Stasis Level of Fact

Unlike the regulatory agencies, the question at the bottom of stasis (Are there health risks related to HVHF?) is answered with a firm yes (fig. 9.7). Southwest Pennsylvania Environmental Health Project not only asserts that there are health risks; it also notes that the risks are "substantial."

Stasis Level of Definition

Unlike the other sites that limited discussion of health issues related to HVHF, the EHP website links to emerging research on intersecting types

---

Data suggest that unconventional oil and gas development (UOGD or "fracking") poses substantial risk to public health from the toxic chemicals used during the industrial process, as well as from other associated problems. It is believed that the potential consequences of "fracking" begin at the onset of drilling and may last long after the operation has ended. (Southwest PA Environmental Health Project, 2021, Health Issues).

Figure 9.7. Southwest PA Environmental Health Project, fact.

of health risks of HVHF, clearly offering answers to the definitional level of stasis (What are the health risks?). The website contains links to academic, community, and nonprofit studies about HVHF environmental health risks that enable an informed discussion in response to this question (fig. 9.8).

Definitions of risk are more specific than in the previous regulatory and advisory institutional websites, where risks were framed as "uncertain" and "monitored" by institutional experts. Instead, the Southwest PA Environmental Health Project includes research about how air quality, water quality, noise and light, stress, soil, and harm to human bodies can be caused by HVHF.

Also, health risks are delineated by their short-term and long-term effects. For example, the "air quality" section includes details about how the different parts of HVHF contribute to issues of human health (fig. 9.9).

The accompanying research, found throughout the site, backs the assertions for each area of concern and references research projects conducted by various institutions with whom EHP has collaborated.

Research is mounting on the emissions from UOGD at all stages and on health effects experienced by nearby residents. Many of the toxic chemicals that have been found in air and water samples around UOGD operations have well known adverse health effects. For example, benzene is a known carcinogen (cancer-causing), toluene is a neurotoxin, and hydrogen sulfide irritates the lungs and can cause asthma. Noise and light pollution is associated with hearing loss, sleeplessness, and other health issues. (Southwest PA Environmental Health Project, 2021, Health Issues).

Figure 9.8. Southwest PA Environmental Health Project, definition.

People who are exposed to high levels of polluted air from UOGD ("fracking") can experience many health effects, and children are more at risk than adults. Researchers report that some health problems associated with breathing bad air can happen right away:
- Burning eyes
- Sore throat
- Stomach pain and nausea
- Headaches
- Tingling/numbness in extremities

Researchers are also studying other long-term effects from air pollution such as cancer, cardiovascular disease, and poor birth outcomes. (Air Quality).

Figure 9.9. Southwest PA Environmental Health Project, definition.

Stasis Levels of Value and Jurisdiction

There are several jurisdiction suggestions based on the definitional level of stasis that answer the questions: Should there be action based on health risks? What is the action suggested based on what authority? The Southwest PA Environmental Health project offers many *ought* suggestions based on values of environmental justice and "green politics." For example, for each area of environmental concern (figs. 9.10a, 9.10b), there are short-

Did you experience a concerning
event, possibly caused by nearby oil
and gas activity? Use this form to
report your experience.

Figure 9.10a. Southwest PA Environmental Health Project, definition.

Figure 9.10b. Southwest PA Environmental Health Project, definition.

term and long-term suggestions so citizens can self-monitor for the area of concern, creating a nuanced and "layered" understanding of risk and promoting agency for public participants to contribute to risk reporting. Also, participants are actively asked to report risks, inviting stakeholder knowledge into the risk-defining process.

The links, such as the "Air Quality" option, offer instructions about the use of simple tools such as Purple Air monitors that link to a larger community science project to monitor air quality rather than waiting for an expert to do the work. The site also offers practical steps that stakeholders can take at home to reduce air quality concerns as well as linking to resources about specific health concerns. Finally, stakeholders can link to legal advice; to other nonprofit community groups researching environmental health risks related to HVHF; to environmental reporting links for potential hazards such as spills or other violations; and to emergency planning tips, including directions for filing a right-to-know request and instructions about attending Local Emergency Planning Committee meetings (figs. 9.11a, 9.11b).

The website also contains a section for healthcare providers (fig. 9.12), including continuing education for those working in HVHF industries and links to professional groups that routinely investigate the link between HVHF and public health concerns.

The site implements a green politics ideology that values intersectional environmental justice principles and enables agency for individual stakeholders living near HVHF activity. Nature is seen as a complex ecosystem and complex interaction between humans and the natural world. Table 9.2 summarizes this discussion of the three websites from a stasis theory perspective, showing how the *is* of health risk leads to *ought* actions.

While the Southwest PA Environmental Health Project website offers much more transparency in discussions about health risks of HVHF, and these deliberative moves should be adopted by other agencies, the lack of precautionary language toward an industry that clearly has environmental and health risks becomes: How can marginalized spaces, especially those with a history of environmental injustice and now faced with another set of environmental and human health disparities, adopt a precautionary form of deliberation to enact justice in action?

Conclusion

The Southwestern PA Environmental Health Project has made great strides to offer support and opportunities for action to local stakeholders

Figure 9.11a. Southwest PA Environmental Health Project, definition.

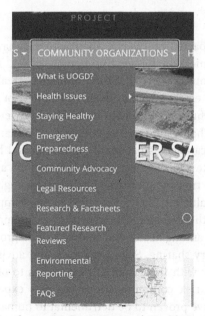

Figure 9.11b. Southwest PA Environmental Health Project, definition.

Figure 9.12. Southwest PA Environmental Health Project, definition.

who might be impacted by HVHF. Their activities have implications for technical communication for environmental action. Technical communication in the Southwestern PA Environmental Health Project reveals intersectional and environmental ideologies as part of the lower levels of stasis (fact and definition) that help to guide upper levels of stasis (value and jurisdiction) that result in policy or action. Similarly, this site models how participants can access pertinent environmental health information through immediate access to actionable suggestions and be part of risk reporting in more transparent ways. This is increasingly important in the realm of public health and likely of comfort to those impacted by the many layered "risks" in the industry. The website also contains more clear links between expertise among health and environmental groups and is a model for interdisciplinary sharing. State regulatory and advisory groups would do well to partner with these community groups to offer more nuanced and participatory risk communications beyond "expert" monitoring of many risks that have proven to be detrimental to human health. Similarly,

Table 9.2. Agency versus Nonprofit Health Risk Representations

	Pennsylvania Department of Environmental Protection	Pennsylvania Department of Health	Southwest Pennsylvania Environmental Health Project
Fact (*Is*) Are there health risks related to HVHF?	Short-term studies on site say no are some health risks	Yes that there are more health risks than previously supposed	Yes
Definition (*Is*) What are the health risks?	Current studies on site do not examine health effects	Unclear	Several risk factors identified and explained
Value (*Ought*) Should there be action based on health risks?	Call for more studies	Call for more studies	Studies posted on site call for more action
Jurisdiction (*Ought*) What is the action suggested based on what authority?	Continue industry; manage risks via monitoring	Continue industry; says DEP should manage risks via monitoring (PA DOH says it is advisory only)	Strategies for identified risk factors; calls for industry change, but no outright call for industry ceasing as part of advocacy; reactionary vs. precautionary
	Storyline: Administrative Rationalism	Storyline: Administrative Rationalism	Storyline: Green Politics

a more robust technical communication mechanism should be put into place to avoid the "stalling" of action due to the lag time for studies of health risk. Successful health agency partnerships, such as those seen in New York State, could be explored more broadly prior to industry activity, particularly for marginalized areas. If industry commences, there should be more pathways for those who experience health effects to report their experiences in a meaningful manner.

However, environmental technical communicators should also consider how better to involve "precautionary" language for industry when outcomes are unknown, especially as it applies to emerging health studies that are beginning to tell a story of many interconnected health risks. While it is laudable that Pennsylvania state funds will be used to explore HVHF human risks through public health studies, these have only come about through outcry by concerned citizens and concerned health professionals. Indeed, questions remain about where the burden of proof lies for determining HVHF's risks, and that leads to a consideration of environmental justice and intersectional and precautionary approaches to technical communication.

In the end, groups like the Southwest Pennsylvania Environmental Health Project often must take a reactive stance, offering "tools" for concerned citizens only after the industry has already been in place for some time. These community health groups may consider, as the next point of advocacy, more actively using health research to disrupt the HVHF industry before more harm occurs to human bodies. Indeed, state environmental and health agencies should consider why nonprofits and publics should have to bear the burden of proof within deliberation; state regulatory agencies should intentionally adopt precautionary language as part of their routine discourse. This ideological shift will lead to a shift in policy and, possibly, abandoning HVHF in Pennsylvania as has happened in New York.

Note

1. The website for the Southwest Pennsylvania Environmental Health Project underwent an extensive revision between the time of writing this chapter and publication. Consequently, the visuals here represent an older iteration of the website. The organization's missions, values, and activities remain the same, however.

References

Black, B., & Ladson, M. (2012). The legacy of extraction: Reading patterns and ethics in Pennsylvania's landscape of energy. *Environmental Histories of the Mid-Atlantic, 79*(4), 377–394.

Brady, W. (2011). *Hydraulic fracturing regulation in the United States: The laissez-faire approach of the federal government and varying state regulations.* University of Denver, Sturm College of Law.

Colaneri, K. (2016). Former state employees say they were silenced on drilling. *State Impact, Pennsylvania.* June 19. Retrieved from https://stateimpact.npr.org/pennsylvania/2014/06/19/former-state-health-employees-say-they-were-silenced-on-drilling/

Concerned Health Professionals of New York. (2020). *Compendium of scientific, medical, and media findings demonstrating risks and harms of fracking (unconventional gas and oil extraction).* Retrieved from https://concerned-healthny.org/

Druschke, C. G. (2014). With whom do we speak? Building transdisciplinary collaborations in the rhetoric of science. *Poroi, 10*(1), 1–7.

Dryzek, J. S. (2005). *The politics of the earth: Environmental discourses.* Oxford University Press.

Fahnestock, J., & Secor, M. (1985). Toward a modern version of stasis. In C. W. Kneupper (Ed.), *Oldspeak/Newspeak* (pp. 217–226). Rhetoric Society of America.

Frost, E. (2015). Apparent feminism as a methodology for technical communication and rhetoric. *Journal of Business and Technical Communication, 30*(1), 3–28.

Funtowicz, S. O., & Ravetz, J. R. (1993). Science for the post-normal age. *Futures, 25*(7), 739–755.

George, B. (2019a). Communicating activist roles and tools in complex energy deliberation. *Communication Design Quarterly, 7*(1), 40–53. https://dl.acm.org/doi/10.1145/3331558.3331562

George, B. (2019b). Language and environmental justice: Articulating intersectionality within energy policy deliberations. *Environmental Sociology, 5*(2), 149–162. https://www.tandfonline.com/doi/full/10.1080/23251042.2019.1605958

Goggin, P. (2009). *Rhetorics, literacies, and narratives of sustainability.* Routledge.

Grabill, J., & Simmons, M. (1998). Towards a civic rhetoric for technologically and scientifically complex places: Invention, performance and participation. *Technical Communications Quarterly, 7*(4), 415–441.

Hardy, C. (2012). Using the environmental history of the commonwealth to enhance Pennsylvania and U.S. history courses. *Pennsylvania History: A Journal of Mid-Atlantic Studies, 79*(4), 473–494.

Herndl, C. G., & Cutlip, L. L. (2013). How can we act? A praxiographical program for the rhetoric of technology, science, and medicine. *Poroi, 9*(1), 1–13.

Holifield, R., Porter, M., & Walker, G. (2010). *Spaces of environmental justice.* Wiley-Blackwell.

McGreavy, D., Silka, B., & Hart, D. (2012). Creating a place for environmental communication research in sustainability science. *Environmental Communication, 6*(1), 23–43.

Nixon, R. (2011). *Slow violence and the environmentalism of the poor.* Harvard University Press.

Pennsylvania Department of Environmental Protection (PA DEP). (2021). Mission statement. Retrieved from https://www.dep.pa.gov/About/Pages/default.aspx

Pennsylvania Department of Environmental Protection (PA DEP). (2012). *Long-term ambient air monitoring project near permanent Marcellus Shale gas facilities.* Retrieved from http://files.dep.state.pa.us/Air/AirQuality/AQPortalFiles/Monitoring%20Topics/Toxic%20Pollutants/Docs/Long-Term_Marcellus_Ambient_Air_Monitoring_Project-Protocol_for_Web_2012-07-23.pdf

Pennsylvania Department of Health (PA DOH). (2021). About the department of health. Retrieved from https://www.health.pa.gov/About/Pages/About.aspx.

Pennsylvania Department of Health (PA DOH). (2020). Bureau of Epidemiology, Division of Community Epidemiology. *Ewing's Family of Tumors, Childhood Cancer and Total Cancer Standard Incidence Ratio Results for Washington, Fayette, Greene and Westmorland Counties in Pennsylvania.* Retrieved from https://www.health.pa.gov/topics/Documents/Environmental%20Health/Ewings%20Tumors%20SW%204%20Counties.pdf

Phillips, S. (2016). PA supreme court rules with environmentalists over remaining issues in Act 13. *State Impact, Pennsylvania.* September 28. https://stateimpact.npr.org/pennsylvania/2016/09/28/pa-supreme-court-rules-with-environmentalists-over-remaining-issues-in-act-13/

Schiappa, E. (1998). *Constructing reality through definitions: The politics of meaning.* The Center for Interdisciplinary Studies of Writing. University of Minnesota.

Simmons, M. (2007). *Participation and power: Civic discourse in environmental policy decisions.* State University of New York Press.

Southwest Pennsylvania Environmental Health Project. (2021). Retrieved from http://www.environmentalhealthproject.org/

Tanenbaum, M. (2019). Gov. Wolfe funds two studies on fracking health impacts, possible cancer link in Pennsylvania. November 22. *Philly Voice.* https://www.phillyvoice.com/fracking-pennsylvania-health-impact-studies-children-cancer-asthma-tom-wolf/

Walker, G. (2010). Environmental justice, impact assessment and the politics of knowledge: The implications of assessing the social distribution of environmental outcomes. *Environmental Impact Assessment Review, 30*(5), 312–318.

Chapter 10

Participatory Policy

Enacting Technical Communication
for a Shared Water Future

JOSEPHINE WALWEMA

In early 2018, the City of Cape Town (CoCT) was on the verge of running out of water. A combination of low rainfall resulting in falling dam water storage, farming/industry needs, and a population growth that exerted pressure on the existing water system created a doomsday scenario dubbed "Day Zero"—the day the city would be without water. Through a set of actions orchestrated by city government and the residents of Cape Town, Day Zero was averted (see Walwema, 2021; Shepherd, 2019; Borofsky, 2019). This set of innovative measures motivated households and businesses to dramatically reduce high water use. However, the City of Cape Town understood that its fate could not rest in those apocalyptic measures alone. It therefore followed up on those measures with extensive and ongoing discussions between city officials and the public to achieve a secure water future. This chapter examines the resulting policy documents and the contexts from which they emerged to understand the role of technical communication in environmental public policy.

Background

The City of Cape Town (CoCT) lies in the southwestern tip of South Africa where it borders the Atlantic Ocean on the west and the Indian Ocean on the south. This location gives the city its temperate Mediterranean-like climate that brings with it hot, dry summers and cold, often below freezing, winters inland. Its precipitation is characterized by three dominant rainfall seasons from winter to late summer and extreme rainfall in some areas (Gasson, 1998). Thus, the city relies on rainfall to replenish its reservoirs and lakes and on runoff from nearby Table Mountain to supply a bulwark of rivers and canals that date back to the city's founding. While these water sources were instrumental to its founding, the city's growing water consumption rate of 4.7 percent per annum has become unsustainable (C40 Cities, n.d.). Further, the increasing aridification of South Africa has rendered the seasonal rainfall scarce and unevenly distributed. Thus, Cape Town, like other global cities, faces a confluence of water challenges that stem from urbanization, population growth, and agriculture, not to mention the effects of climate change.

The demand for water impacts its allocation and management, however well intentioned the city's policies are. For example, records indicate that most water is consumed by urban dwellers and agriculture at 70 and 30 percent, respectively (Shepherd, 2019). Further, water consumption in Cape Town is colored by the inequities of apartheid era policies, which the city addressed by pricing wealthier consumers at a higher rate in order to subsidize low-income households with up to 6,000 liters of free water to users at communal faucets (Shepherd, 2019).

Following Day Zero, city authorities trained their efforts on policies that would enable city residents to become proactive water users who are sensitive to water complexity considering climate change. The city's policy approach appears to eschew the one-way Jeffersonian approach of experts speaking to the public. It recognized both its obligation to dispense technical knowledge, for example the risk of flooding in low-lying areas, and the public's role in mitigating that risk. Thus, the CoCT has sought to bring together science and local knowledge of "participants in environmentally affected communities" (Wadell, 1990, p. 45). This shift, it is hoped, can best mobilize and engage the public in supporting the city's efforts to keep water flowing. Importantly, the city recognizes that water management requires a comprehensive approach that transcends municipal and certainly individual water use and that it involves all stakeholders.

One such initiative, Water Sensitive Design (WSD), is a response to the "linear design approach," where decisions were solely in the hands of only city authorities whose solutions were technocratic and fragmented. WSD counters that approach by being capacious in its objective to sustainably manage water by "transitioning from linear technology-centered and resource-intensive approaches to cyclical human-centered systemic approaches" (University of Cape Town, para. 1; Madonsela et al., 2019). The approach also informs CoCT's transition to a Water Sensitive City, reflected in the city's water policy initiatives published in 2019–2020 and the basis of this study.

Through critical discourse analysis, I will seek to answer this question: What technical communication strategies frame the policies outlined by the CoCT to address water precarity and bring about a secure water future? The way these actions function as technical communication is of vital importance to our collective understanding of building water sustainability solutions in a warming climate.

Theoretical Framework

Environmental communication has historically rested on appeals to logic to explain complex information. Known as "technical rationality" (Kinsella, 2004), this appeal to technical knowledge entails adjusting the "complexities of the sciences to the intellectual limitations of non-scientific publics" (Gross, 1994, p. 19). Separating scientific from nonscientific publics has been labeled a "deficit model" of public understanding of scientific knowledge—the assumption that publics are ignorant and need to be educated by technical experts. Critics of the deficit model say it ignores the very essence of publics' contexts of knowledge, experience, and relationship to their respective environments. It also sidesteps the multiple means of knowledge making from which inferences are made and conclusions drawn (see Besley & Nisbet, 2013).

Typically, scientists process empirical data, draw conclusions, and share implications with the public. They, however, err by expecting publics to process information similarly (Besley & Nisbet, 2013). This deficit model of communication, scholars have found, is not sufficient to change minds, given that it quite often simply proceeds with the assumption that the technical content is objectively self-evident. While it might create consensus, self-evident content does not necessarily lead to "conviction"

(Waddell, 1990, p. 351). Yet, because technical communication is, for the most part, interdisciplinary in nature and audience centered, it should not discount members of the public whose lived experiences in these environments inform their knowledge of weather patterns even when recounted in lay discourses, nor dismiss them as disinterested bystanders (Hanson-Easey et al., 2015). The public draws from their embedded personal experience and invocation of localized cultural practices to knowledgeably speak about environmental issues. Thus, technical communicators should be motivated to engage the public to deliberate and, hopefully, accept expert-derived public policy.

Building trust necessitates a dialogic interaction that assures that the public *learns* the science and how it affects their lives and that the technical experts *control* the messaging with accurate science. It means acknowledging that risk is socially constructed and shaped by public, political, social, and economic perspectives (Grabill & Simmons, 1998). Thus, solutions to risk require broad public participation at local, national, and international levels if they are to succeed. And to obtain public participation is to interact directly with the public and to solicit their experiential knowledge. This in turn allows the public to adapt those solutions to their local contexts. Yet, as some studies show (e.g., DeCaro & Stokes, 2013), there are limits to stakeholder involvement in environmental decision making.

Most environmental discourse originates in the techno-scientific sphere, and thus warrants techno-scientific terminology, even statistics or data to convey the gravitas of scientific authority (Johnson, 2009, p. 32). Still, it is through civic discourse that technical communication expertise connects with the public it purports to serve. Simmons (2007) found that when experts "decide-announce-defend" their decision making, the public may resort to "strategic, often subversive moves" (p. 48) that undermine the experts. Conversely, Druschke and Hychka (2015) find that the contextual approach leads to broad-based support and fosters ecological success. Breaking down the binary between the technical and the social means inviting technical communication into the sphere of public discourse where it becomes, by necessity, less specialized so the public can "share in the construction of the future" (Goodnight, 1982, p. 232). This recognition "insures the political equality of expert and lay representatives and guarantees closure" (Gross, 1994, p. 17; see also Currie, 2013).

In technical communication, writing about the environment is about bridging the gap between those who create ideas and the public who use them in opportune moments requiring technical expertise. When such important policy matters are left to scientists, industry, and government, they risk excluding the very public needed to effectuate policy. As Simmons (2007) reminds us, lay publics are not synonymous with ignorant publics. Moreover, individuals are constantly managing risk in their lives. What they need is the information to manage those risks. That is how policy translates into action. It is how technical communication shapes and renders technical knowledge valuable, actionable, and effective in bringing about change (Longo, 2000, p. xi).

The lesson here is that because communicating risk information is not simply about interpreting "complex technical arguments" (Waddell, 1990, p. 381), but also about rendering that knowledge actionable in accessible ways, policy should include public participation.

Method

As Feindt and Oels (2005) observe, the environment is about nature, but its problems are socially constructed and debated. Thus, this study uses critical discourse analysis (CDA) (Fairclough 2003, 2010) to examine the CoCT policy documents to answer the question: What technical communication strategies frame the policies outlined by the CoCT to address water precarity and bring about a secure water future? Because CDA attends to the circulation of power and ideology in discourse, texts are merely evidence of existing discourse, which frames "ways of constituting knowledge" and "systematically form the objects of which they speak" (Foucault, 2002, p. 54).

Huckin et al. (2012) define CDA as an "interdisciplinary approach to textual study that aims to explicate abuses of power promoted . . . by analyzing linguistic/semiotic details in light of the larger social and political contexts in which those texts circulate" (p. 107). Similarly, Fairclough et al. (2011) find that CDA is "especially appropriate to a textual analysis working towards an understanding of the complex interrelationships between communication, policy, policy-making processes, and underlying ideological assumptions of and about the people involved in both" (p. 357). Moreover, policy documents are a result of political processes that can legitimize existing structures while masquerading as a vision for the future. I find CDA applicable in this study because it goes beyond the content of the policy texts to inform the technical communication processes in which these texts are framed and function.

Selection of the Texts

Between February 2019 and March 2020, the CoCT published policy and strategic documents outlining Cape Town's water strategy. These documents, housed on the CoCT's website, outline regulatory measures, courses of action, and priorities concerning water use. They are

- *Our Shared Future: Cape Town's Water Strategy* (2020a)

- *Cape Town Water Strategy: Never Waste a Good Crisis* (2019)

- *Status of Water Supply Augmentation and Diversification Projects* (2020b)

These documents, all published by the City of Cape Town, are only differentiated by their respective titles. My analysis focuses on the first two.

At their core, these policy documents are regulatory processes and procedures shaped by technical expertise and situational awareness. They lay out the trajectory of water supply for the City of Cape Town and outline five processes (in the form of commitments) that would result in a secure water future. As Miriam Williams (2009, 2021) has shown, regulatory texts are by nature technical communication, and public policy texts are themselves "sites of TPC research" because they entail a complex writing process between science and technology (Williams, 2021, p. 2). A 2008 *Technical Communication Quarterly* special issue situated public policy and science within TC as did a special issue of *Technical Communication* on election technologies (Gibson, 2008; Dorpenyo & Agboka, 2018). To secure a water future for the City of Cape Town, the solution is encoded in the commitments outlined in the policy documents.

Analysis

I adopted a tripartite analytical procedure that (a) characterized the structure, discourse, and genre of the texts, (b) examined the content, and (c) analyzed the emerging narrative associated with securing a water future for the City of Cape Town.

Structure, Discourse, and Genre of Texts

Cape Town Water Strategy is written in a presentation style genre that integrates technical expertise with the social, cultural, and ethical contexts of water use in the City of Cape Town. It employs complex multimodal semiotics with well-chosen graphics that illustrate how various demographics use water—be it indoor plumbing, communal faucets, or farm water. This choice immediately assumes an audience, as figure 10.1 shows, in depicting the demographics of water users in Cape Town. The invocation of visual details creates what Mando (2016) has called "vicarious proximity" (p. 5), which has the effect of making environmental discourse more apparent. The eye is drawn to visualizing specific places, sites, and locations that are commonplace to most city dwellers and thus analogous to their own water use situations. Such specificity locates the issue close to home.

Visual vivid design serves a functional and informative purpose with content that reinvigorates the narrative around water use. Still, it is difficult to perceive whether presenting different demographics on the same level signifies a vision of equality because while these images foreground a cross section of water users in Cape Town, they also display "how social and political inequalities are manifested in and reproduced through discourse" (Wooffitt, 2005, p. 137). Context matters here because as we learn from Fairclough (2010) unequal power relations manifest through discursive practices, as is evident in these visuals.

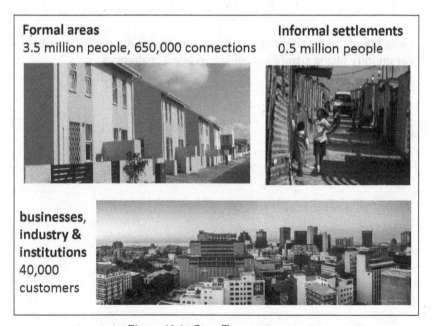

Figure 10.1. Cape Town water use.

Cape Town Water Strategy begins by contextualizing Cape Town's water challenges on a global map that lists major global cities facing water precarity. Figure 10.2 places Cape Town alongside cities experiencing water scarcity on a global map.

Technical communicators have used a variety of visual strategies to inform and persuade audiences (Kostlenick, 2016). While visuals "inherently privilege the viewpoints of the powerful" (Kimball, 2006, p. 379; Welhausen, 2015), maps, especially, offer a snapshot of a specific historical moment in time. Listing Los Angeles, São Paulo, Melbourne, Chennai, Moscow, and Seoul (to mention a few) signals both a rhetorical and technical move. Rhetorically, Cape Town is placed on the world map and at par with other global cities, while technically intimating that Cape Town is not alone in facing water challenges. As a technical communication strategy, it elevates the immediate threat of water precarity the city faces. This kind of content lies somewhere between science and public policy even as it privileges scientific evidence, which potentially undermines its locality and thus relevance and applicability.

The document's content outlines the "core goal" of the policy to "provide reliable, safe, and affordable services to our 4 million customers" (p. 3). This discourse reveals distinct actors grouped into formal/informal settlements

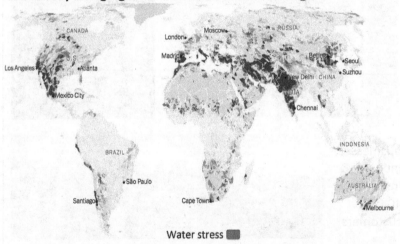

Cape Town's water challenges are not unique, we are part of a global community facing significant climate-related challenges

Water stress ▇

Figure 10.2. Cape Town water challenge on a global scale.

and lumps together business, industry, and institutions (see figure 10.1). Both the text and the discourse convey the composition of this city and the competing factors at play to secure specific interests. In essence, each category of residents is expected to stay in their respective lanes—communal, farm, and industrial water users. A chart depicting how "free flowing reservoirs" from 1834 to 1997 led to dam construction during the formation of Cape Town as a municipality functions as a technological and historical signifier with periods of human intervention to keep the water flowing. Showing water supply, from its sources in dams and fountains to locations in homes and pools compels the public to see water in their faucets as part of wider ecological system and that therefore warrants understanding.

The pattern is repeated in *Our Shared Future* in which we see a chronology Cape Town's water resources development (fig. 10.3).

This chronology, while consistent with technical communication's attentiveness to public accounting of knowledge, uses evidentiary knowledge to help the public visualize water-related data at a granular level. Details such as "dam construction," and "water rationing," remind the public that throughout the city's water history, humans have intervened to construct, ration, and control its water supply. From the magnificent Table Mountain, to aquifers, dams, and hydraulic systems, Cape Town has found ways to bring water to farms, townships, homes, and businesses through people and technology. Therefore, effectively conveying technical and actionable information, and framing the five commitments shown in figure 10.4, is not a pipe dream.

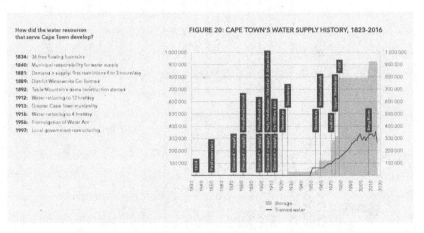

Figure 10.3. Cape Town water history from 1834 to 1997.

1

SAFE ACCESS TO WATER AND SANITATION

The City of Cape Town metropolitan municipality[2] will work hard to provide and facilitate safe access to water and sanitation for all of its residents in terms of well-defined minimum standards. In particular, the City will work with communities in informal settlements and with other stakeholders to improve the daily experience of access to water and sanitation, with an emphasis on building trust and increasing safety within these communities through this process.

2

WISE USE

The City will promote the wise use of water by all water users. This will include promoting water conservation behaviour through (a) pricing water with reference to the cost of providing additional supply, while retaining the commitment to provide a basic amount of water for free for those not able to afford this; (b) revising by-laws and planning requirements, and using other incentives to support water efficiency and the treatment and reuse of water; (c) supporting active citizenship by substantially improving customer management and engagement; and (d) managing the water network effectively to reduce losses and non-revenue water.

3

SUFFICIENT, RELIABLE WATER FROM DIVERSE SOURCES

The City will develop new, diverse supplies of water including groundwater, water reuse and desalinated water, cost effectively and timeously to increase resilience[3] and substantially reduce the likelihood of severe water restrictions in future. The City is committed to increasing supply by building affordable new capacity of approximately 300 million litres per day over the next ten years, and in suitable increments thereafter, in a way that is adaptable and robust to changes in circumstances.

4

SHARED BENEFITS FROM REGIONAL WATER RESOURCES

The City will work with key stakeholders and partners, including other urban and agriculture water users and other spheres of government, to make the most of the opportunities to optimise the economic, social and ecological benefits of regional[4] water resources, and to reduce the risks. The City will do this through collaborative processes.

5

A WATER-SENSITIVE CITY[5]

The City will actively facilitate the transition of Cape Town over time into a water-sensitive city with diverse water resources, diversified infrastructure and one that makes optimal use of stormwater and urban waterways for the purposes of flood control, aquifer recharge, water reuse and recreation, and is based on sound ecological principles. This will be done through new incentives and regulatory mechanisms as well as through the way the City invests in new infrastructure.

Figure 10.4. The city's water commitments: *Our Shared Future.*

The commitments are framed in plain language like "safe," "access," "wise water use," "responsible use of rainwater, greywater and groundwater from private boreholes and well points for non-drinking purposes"—language that reflects South African usage (p. 36). Listing key actions like "supporting active citizenship" echoes the intervention rhetoric of the Day Zero crisis. The policies are framed as commitments whose effect is to regulate human behavior (see Williams, 2009, p. 451).

The Content

The content outlines a time frame to implement the strategy over a 20-year commitment that begins with capacity building (three years), water resilience (10 years), and developing water sensitivity (seven years) with the remaining slides devoted to each one of the five commitments also featured in the second strategy document, *Our Shared Future*. By making comparisons between the city of Perth in Australia and Cape Town, both of which "rely on winter cold fronts for their rain" (p. 7), readers are pointed toward a future Cape Town might inhabit. For example, the city of Perth now has "ground water and desalination" as sources of water (p. 8). This gesture toward similarity is a deliberate move that simultaneously creates affinity and similarity with that city in countering water shortage.

A graphic attributed to World Bank data (fig. 10.5) is an intertextual source for policymaking to promote the public's contextual knowledge by familiarizing them with comparable global data. Moreover, articulating a water problem beyond local needs allows the public to grapple with myriad discourses and meanings related to changing weather patterns and how current and potential developments may affect their local communities. An integral component of Fairclough's methodology is to situate texts within a wider context or what he terms "social practice" (Fairclough, 2010, p. 164). For example, the documents revisit the water history of Cape Town, name the sources of water the city has historically relied upon, and indicate the limitations of those water sources while laying out other sources of water (like comparable cities have done). Such discourse is both universal in visualizing the contextual knowledge that informs the policy documents and particular in applying it to households.

In publicizing threats to the water supply over the years, this content renders historical knowledge present and explicit. As the Brundtland (1985) report has made apparent, the world faces a threatened future and water shortage renders that future all the more precarious. Thus, in

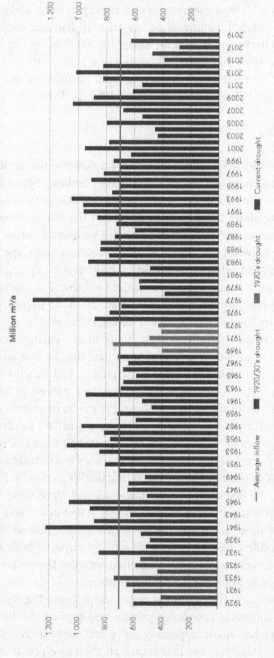

Figure 10.5. Annual flow of water into Cape Town's dams.

these documents, water is imbued with a kind of presence that makes it salient so that the image's effect persuades the public to consider its own solutions to this local problem. This strategy has the effect of activating the public as direct observers of changing water patterns.

Our Shared Future: Cape Town's Water Strategy lays out the city's "new relationship with water" (p. 2). This document was "published for public comment in March 2019" and reportedly attracted "about 38,000 comments during the public participation process for a proposed water by-law amendment" (Mortlock, 2018, para. 1). This dialogic participation is emblematic of the contextual model that might engender shared actions for sustainability. Our Shared Future is characterized as "a high-level strategy document that sets out the city's approach to water, identifies key priorities, and articulates a set of core commitments" (p. 3) that indicate a coordinated approach to policy dissemination (fig. 10.6). The five commitments co-opt the possessive "our" to negotiate interactional meaning by engaging with readers as insiders. "Our" collapses the identity of city dwellers with that of the city and intimates that the city's fortunes are just as equally appropriated. This, of course, is a far cry from the income inequality of Cape Town.

Like Cape Town Water Strategy, Our Shared Future is interspersed with technical and ethnographic data on water, its use, and its sources. Such content deemphasizes abstract notions of sustainability, resilience, and inclusion by fleshing out the implications of each in pragmatic and

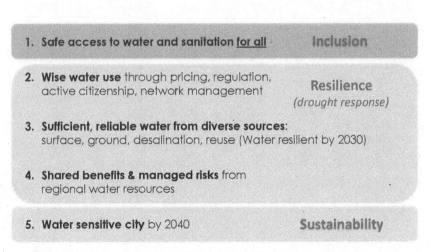

Figure 10.6. Five commitments in the water strategy.

concrete terms. The language is aspirational and could be read as constitutive in its effort to visualize a secure water future. Such language creates a shared sense of value reflecting the active role of discourse in constructing a possible future (Foucault, 2002).

A map depicting the Cape Town water supply system visualizes the various sources of water and how they feed into the water supply shown in figure 10.7.

This map reflects the city's commitment to reducing regional water resource risks, serves as a primary source of knowledge, and functions as a cognitive model that compels the public to visualize the water supply. Williams (2009) labels this kind of discourse a rulemaking process designed to increase trust. On the whole, it exemplifies how discourse can provide "cohesion" without any significant content (Fairclough, 1992, p. 77). From reading and interpreting this map, the public is now better able to consider why, for example, "clearing alien invasive vegetation" can help increase water flow. In technical communication, visualization based on the gestalt principles of symmetry and proximity, and good continuation, account for

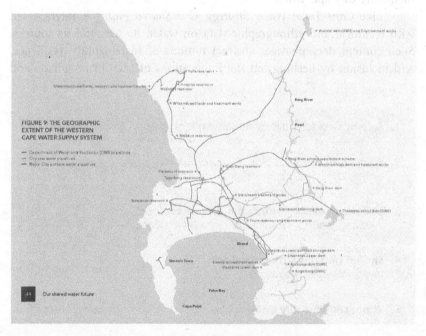

Figure 10.7. Western Cape Water supply system.

how people recognize patterns to appropriate technical information. Such an element helps the public put disparate pieces of information together and create a deeper understanding of the issue at hand. The language of inclusion, resilience, and sustainability signals a form of action that cues readers to recognize that public participation creates cooperation among government and citizens.

EMERGING NARRATIVE

Fairclough and Wodak (1997) argue that "discourse analysis is interpretative and explanatory" (p. 280). While this is true, inevitably, policy is where multiple agendas are enacted. Thus, we see in these policy documents a plethora of data and facts that project institutional authority and give policymakers cover to propose bold measures. While they help the public develop place-based connections with environment, the analysis shows so much more. The policy discourse addresses social problems, constitutes society and culture, does ideological work, is historical, and mediates the text as a blueprint for attaining a secure water future. Considering how integral water policy was to the city's formation (fig. 10.3), not being mindful of that history nearly broke the city—as Day Zero illustrated.

Water conservation efforts are tackled within an ecosystem that includes clearing vegetation in catchment areas (CoCT, 2020a) and bringing plumbing into informal settlements, which improves sanitation and integrates all residents into the city's water infrastructure. It's worth noting that residents in informal settlements have historically survived on the margins of the city and have not been afforded services. Granted, the apartheid era ideology has much to blame for this neglect. However, that ideology has been sustained by the wealth and income inequality that pervades Cape Town. That changes with this inclusive vision for a shared future.

Although visibly informed by scientific methods, data, and measurements, technical communication strategies assure deliberation of the choices offered by the science and the collective values of the people of Cape Town. To determine choices about the future, the documents highlight public water needs, examine water supply and sources, and compute associated costs within the themes of sustainability and diversification. In reflecting not just the science but public values, the discourse invigorates public engagement and can function as a persuasive technique for an otherwise apathetic public. This tacit nod to the complexity of crises and the need for a broader societal response is more efficacious than a simple imposition of restrictions "because the science says so."

The scientific aspects of the documents employ ample use of images, line graphs, and tables to show the trajectory of water use, make comparisons with other cities, and demonstrate how to audit the city's own use of water with quantifiable evidence. That strategy assumes a temporal sequence that advances the claim that the water situation can be righted if the public is incentivized to secure a water future free from the brinkmanship of Day Zero. The evidence-based claims conveyed in key language include "reduce runoff and peak flow; storage with potential for reuse; optimal use; capture and reuse stormwater for non-potable purposes." Such wording operates as cues (Fairclough, 1992) that cultivate a sense of trust with the public and leads to an implicit allocation of rights and responsibilities for both the public and city technocrats.

Communicating technical information is not simply a matter of translation. It ought to be based on broad knowledge from science and publics. Thus, the perceived dichotomy between technical experts—who presume that risks are objectively self-evident owing to their scientific bona fides—and the nonscientific publics whose knowledge is experiential seems false.

Implications

This analysis has attempted to show how the CoCT policy documents are, at their core, problem-solving documents. From the analysis of document design and content structure we learn several things. First, *inclusive public participation* brings with it the benefits of deliberation and accountability, which, in the CoCT entailed openly contending with the water crisis that precipitated Day Zero and proposing strategies to move the city toward a more sustainable water future. The discourse orchestrates strategic technical communication practices designed to anticipate and alter public behavior related to water through carefully constructed design choices, which scholars in technical communication have shown can influence users (Redish et al., 2010).

The ideals embedded in the cliché of "never wasting a good crisis" advance big policy ideas derived from the immediacy of the drought and Day Zero. That crisis created an opportunistic moment for (a) policymakers to promulgate new regulatory processes and (b) primed public acceptance of those measures. Seizing the crisis demonstrates how technical communication focuses on communicative events rather than decontextualized phrases (Wodak & Meyer, 2015).

Expert-driven knowledge need not be the only driving force. As this analysis has shown, the Water Sensitive Design (WSD) approach adopted by policymakers is a technical communication strategy that integrates land/water planning and stewardship to sustainably manage water. Because it begins with an audit of the city's water status—challenges, (mis)management, diversification—it logically leads to next steps such as laying out a roadmap for "transitioning from linear technology-centered and resource-intensive approaches to cyclical human-centered systemic approaches" (University of Cape Town, 2021). In terms of a deliberative approach, this shift is important for two reasons: it is *informational* (based on the water status audit) and *collaborative* (in its efforts to get public buy-in). Because the linear approach placed the responsibility for water management squarely on the shoulders of city government, it meant that the public was oblivious to the technicalities of water supply and only concerned themselves with individual households' or companies' water bills. It invokes public involvement in getting the city to water-sensitive status through accounting for the entire water cycle and the various stakeholders represented in the city's urban-cultural and socioeconomic divide.

This practice has reshaped the public's relationship to water. On the *Wise Water* use and diversification front, the public has been empowered to engage in the widespread harvesting of stormwater and reuse of gray water (generated from bathroom sinks, showers, tubs, and washing machines) to flush toilets and replenish landscapes. Both these sources boost the city's potable water, reduce pressure on the water system, and cultivate a "water-sensitive city."

Scientific epistemology is infused in the policy documents as they chronicle the environmental pressures that exacerbate the water situation without technical jargon. The result is public awareness of the causes of water scarcity such as paucity of fresh and ground water, flood risks from sea-level rise, heat risk from the semi-arid location of Cape Town, and high water use by Capetonians. As if these challenges were not enough, the city recognizes the social and financial pressures of urbanization and poverty such as residents of low-lying parts of the city being prone to floods. Making this knowledge explicit has resulted in inclusive planning and promulgated community practical policies that alleviate the worst elements of flooding like raising flooring of homes in shanty towns so they no longer sit in the flooded wetland area, installing drainage to channel flood water away from homes, and clearing drainage systems of litter. In engaging this type of discourse, the CoCT legitimates the power of expert knowledge but renders it practical knowledge for the residents of those areas.

Ideas are not expressed as hypothetical, but they are expressed as provisions. This purposive use of language not only facilitates action but helps shape attitudes that predispose Cape Town residents toward practical engagement in securing a water future for their city. Residents see not just their rights but their responsibility to a secure water future. Stakeholder engagement assures that involvement at all steps affirms that science is relevant, accessible, circulated, and impactful as it ties in with people's lived experience.

Decentralizing the water management system through a water-sensitive design has elements of distributing responsibilities for all stakeholders. This is an element of participatory risk technical communication. The CoCT can buffer the impacts of climate change by clearing "invasive alien plants in the catchments of the major dams," diversifying water sources, and rehabilitating degraded waterways, but the public has to adopt a water-sensitive mindset and be willing to behave differently. For example, individual households can fix leaky faucets, communities can clear foliage from drainage systems in their immediate locales, and families can proactively limit flooding by constructing their houses with susceptibility for flooding in mind.

Because the policy documents are not limited to outreach and educating the public but reflect a participatory acceptance of responsibilities, they are not a call for passive acceptance. Oriented toward the future, they are rife with imaginative solutions associated with water security. This advocacy for action renders the documents practical technical communication for environmental action and for understanding the environment as an ecosystem within which lives are lived and livelihoods made.

A negotiated approach to public policy that acknowledges the social construction of risk allows for socially constructed solutions that draw from the public's political, social, economic, and psychological perspectives. Interacting more directly with the public and soliciting their collective wisdom is beneficial and can empower the public to participate by "curat(ing) their own risk information experience" (Welhausen, 2017, p. 52). A combination of technical and experiential knowledge can lead to a wider distribution of shared knowledge through both institutional and extra-institutional channels. It means not assuming that the public is a cultural/social monolith even if they are located in the same city. It does not assume that the public will always take scientific knowledge at face value even in moments of crisis.

Conclusion

In this chapter I have used critical discourse analysis to explore the discourse and technical communication strategies surrounding the City of Cape Town's policy strategy to attain a secure water future. About three years after the near apocalypse that was Day Zero, Cape Town recognized that reliable water supply was no longer a given. While the city cannot prevent aridification, it can design a water policy that brings people and technology together to manage the water it has even as it harnesses technologies of reuse, rainwater, and desalination. While the policy documents do not make clear the nature of community consultation that precipitated them, we can infer the role of technical experts in encoding them with regulatory discourse that would result in a secure water future. This analysis describes the technical strategies that inform this policy.

The documents delineate the interests of various residents across the economic-sociocultural divide without collapsing the notion of public into one homogeneous heap. This choice strategically illuminates the material and systemic constraints that impact a cross section of the people of Cape Town and attends to the material details of each sector of city residents while recognizing and mobilizing their shared responsibility in securing a water future. The approach might implicitly acknowledge how past water policy, limited as it was to city dwellers and famers while averting its gaze from township dwellers, did little to advance a sustainable water future. This outlook thus, perhaps, signals a new dawn for Cape Town and similarly affected cities.

References

Besley, J. C., & Nisbet, M. C. (2013). How scientists view the media, the public, and the political process. *Public Understanding of Science* 22, 644–659.

Borofsky, Y. (2019, July 15). After Day Zero: How water conservation undermines adaptation in Cape Town. *The Breakthrough Institute*. Retrieved from https://thebreakthrough.org/journal/no-11-summer-2019/after-day-zero

Brundtland, G. H. (1985). World commission on environment and development. *Environmental Policy and Law, 14*(1), 26–30.

C40 Cities. (n.d.). *Cape Town: Water conservation and demand management (WCWDM) programme.* https://www.c40.org/awards/2015-awards/profiles/64

City of Cape Town (CoCT). (2019). *Cape Town water strategy: Never waste a good crisis.* Retrieved from https://www.capetown.gov.za/Document-centre

City of Cape Town (CoCT). (2020a). *Our shared future: Cape Town's water strategy.* Retrieved from https://www.capetown.gov.za/Document-centre

City of Cape Town (CoCT). (2020b). *Status of Water Supply Augmentation and Diversification Projects* (2020). Retrieved from https://resource.capetown.gov.za/documentcentre

Currie, S. (2013). How to save the world: The ideologies of four ENGOs and their discursive constructions of climate change. Applied linguistics and discourse studies (master's thesis). Carleton University.

DeCaro, D. A., & Stokes, M. K. (2013). Public participation and institutional fit. *Ecology and Society, 18*(4), 40.

Dorpenyo, I., & Agboka, G. (2018). Guest editors' editorial: Election technologies, communication, and civil engagement. *Technical Communication, 65*(4), 349–335.

Druschke, C. G., & Hychka, K. C. (2015). Manager perspectives on communication and public engagement in ecological restoration project success. *Ecology and Society, 20*(1), 58.

Fairclough, N. (1992). Discourse and text: Linguistic and intertextual analysis within discourse analysis. *Discourse and Society, 3*(2), 193–217.

Fairclough, N. (2003). *Analyzing discourse: Textual analysis for social research.* Routledge.

Fairclough, N. (2010). *Critical discourse analysis: The critical study of language,* second edition. Routledge.

Fairclough, N., Mulderrig, J., & Wodak, R. (2011). Critical discourse analysis. In T. Van Dijk (Ed.), *Discourse studies: A multidisciplinary* introduction (pp. 357–378). Sage.

Fairclough, N., & Wodak, R. (1997). Critical discourse analysis. In T. A. Van Dijk (Ed.), *Discourse as social interaction* (pp. 258–284). Sage.

Feindt, P. H., & Oels, A. (2005). Does discourse matter? Discourse analysis in environmental policy making. *Journal of Environmental Policy and Planning, 7*(3), 161–173.

Foucault, M. (2002). *Archaeology of knowledge.* Routledge.

Gasson, B. (1998). The biophysical environment of the Western Cape Province in relation to its economy and settlements. School of Architecture and Planning, University of Cape Town for the Department of Local Government and Housing (Directorate of Development Promotion) of the Province of the Western Cape. Cape Town, South Africa.

Gibson, K. (2008). Guest editor's introduction: Science and public policy. *Technical Communication Quarterly, 18*(1), 1–8.

Goodnight, G. T. (1982). The personal, technical, and public spheres of argument: A speculative inquiry into the art of public deliberation. *Journal of the American Forensic Association, 18*(4), 214–227.

Grabill, J. T., & Simmons, W. M. (1998). Toward a critical rhetoric of risk communication: Producing citizens and the role of technical communicators. *Technical Communication Quarterly, 7*(4), 415–441.

Gross, A. (1994). The roles of rhetoric in the public understanding of science. *Public Understanding of Science, 3*(1), 3–24.

Hanson-Easey, S., Williams, S., Hansen, A., Fogarty, K., & Bi, P. (2015). Speaking of climate change. *Science Communication, 37*(2), 217–239.

Huckin, T., Andrus, J., & Clary-Lemon, J. (2012). Critical discourse analysis and rhetoric and composition. *College Composition and Communication, 64*(1), 107–129.

Johnson, L. (2009). Environmental rhetorics of tempered apocalypticism in *An Inconvenient Truth*. *Rhetoric Review, 28*(1), 29–46.

Kimball, M. A. (2006). London through rose-colored graphics: Visual rhetoric and information graphic design in Charles Booth's maps of London poverty. *Journal of Technical Writing and Communication, 36*(4), 351–381.

Kinsella, W. J. (2004). Public expertise: A foundation for citizen participation in energy and environmental decisions. In S. P. Depoe, J. W. Delicath, & M. A. Elsenbeer (Eds.), *Communication and public participation in environmental decision making* (pp. 83–95). State University of New York Press.

Kostelnick, C. (2016). Mosaics, culture, and rhetorical resiliency: The convoluted genealogy of a data display genre. In M. A. Kimball & C. Kostelnick (Eds.), *Visible numbers: Essays on the history of statistical graphics* (pp. 177–206). Ashgate.

Longo, B. (2000). *Spurious coin: A history of science, management, and technical writing*. State University of New York Press.

Madonsela, B., Koop, S., Van Leeuwen, K., & Carden, K. (2019). Evaluation of water governance processes required to transition towards water sensitive urban design: An indicator assessment approach for the City of Cape Town. *Water, 11*(2), 292. https://doi.org/10.3390/w11020292

Mando, J. (2016). Constructing the vicarious experience of proximity in a Marcellus Shale public hearing. *Environmental Communication, 10*(3), 352–364.

Mortlock, M. (2018, March 17). City of CT to decide on proposed water by-law amendment. *Eyewitness News*, Cape Town. http://ewn.co.za/2018/03/17/city-of-ct-to-decide-on-proposed-water-bylaw-amendment

Redish, J. G., Chisnell, D. E., Laskowski, S. J., & Lowry, S. (2010). Plain language makes a difference when people vote. *Journal of Usability Studies, 5*(3), 81–103.

Shepherd, N. (2019). Making sense of "Day Zero": Slow catastrophes, Anthropo-cene futures, and the story of Cape Town's water crisis. *Water, 11*(9), 1744.

Simmons, M. (2007). *Participation and power: A rhetoric for civic discourse in environmental policy.* State University of New York Press.

University of Cape Town. (2021). Water sensitive design. *Future Water.* http://www.futurewater.uct.ac.za/FW-wsd

Waddell, C. (1990). The role of *pathos* in the decision-making process: A study in the rhetoric of science policy. *Quarterly Journal of Speech, 76*(4), 381–400.

Walwema, J. (2021). Rhetoric and Cape Town's campaign to defeat Day Zero. *Journal of Technical Writing and Communication, 51*(2), 103–136.

Welhausen, C. A. (2017). At your own risk: User-contributed flu maps, participatory surveillance, and an emergent DIY risk assessment ethic. *Communication Design Quarterly Review, 5*(2), 51–61.

Welhausen, C. A. (2015). Power and authority in disease maps: Visualizing medical cartography through yellow fever mapping. *Journal of Business and Technical Communication, 29*(3), 257–283.

Williams, M. F. (2009). Understanding public policy as a technological process. *Journal of Business and Technical Communication, 23*(4), 448–462.

Williams, M. F. (2021). Gun control and gun rights: A conceptual framework for analyzing public policy issues in technical and professional communication. *Technical Communication Quarterly,* ahead of print, 1–11.

Wodak, R., & Meyer, M. (Eds.). (2015). *Methods of critical discourse studies.* Sage.

Wooffitt, R. (2005). *Conversation analysis and discourse analysis: A comparative and critical introduction.* Sage.

Chapter 11

Rhino Crash

Teaching Science, Medical, and
Environmental Writing for Social Action

MICHELLE HALL KELLS

In his essay *It All Turns on Affection*, Wendell Berry weaves an ecological ethic around two values—imagination and appreciation. "For humans to have a responsible relationship to the world, they must imagine their place in it," Berry writes. "As imagination enables sympathy, sympathy enables affection. And in affection we find the possibility of neighborly, kind, and conserving economy" (2015, p. 14). Wendell's ethic offers a productive frame for teaching technical communication for environmental action, reminding us that education is most efficacious when we help students find their way into issues. Deeply troubling environmental issues can move writers and activists to rhetorical action in productive ways through the heart lines of appreciation and imagination. Above all, teaching technical communication for environmental action can make the heart lines to interspecies relationships palpable, visible, and present.

In a political climate overshadowed by a suspicion toward science and a disdain for the liberal arts, the challenge for teachers of environmental rhetoric and technical communication is engaging not only resistance to reason, but a failure of imagination. Bob Hyland aptly argues in chapter 3 of this volume, "In Defense of a Greenspace: Students Discover Agency

267

in the Practice of Community-Engaged Technical Communication," that our students "have been surrounded by a discourse of impending doom concerning the environment." Helping students engage with difficult environmental issues and resist despair is one of the greatest challenges of teaching technical communication for environmental action. This chapter is based on my experience designing and implementing a pedagogy for my ENGL 413 Science, Medical, and Environmental Writing. This project foregrounds the work of ecological stewardship through interspecies kinship ties as a critical feature of environmental education in technical communication.

During the spring of 2020, my students and I invited a local equine veterinarian specialist to partner with us in an international collaborative project. Together we enacted the roles of citizen scholars, technical writers, and environmental activists in support of international rhino conservation efforts. This chapter is organized around a set of 10 "P's in a Pod" to narrate, describe, explicate, and analyze the design, implementation, and delivery of the "Rhino Crash" project. I use these "P's in a Pod" as a heuristic for teaching technical communication for environmental action. My ENGL 413 students and I applied this conceptual framework in a recursive fashion to navigate the challenges of engaging rhino conservation as an environmental case study. Our 10 pedagogical "P's in a Pod" include: 1. The Power of *Positivity*, 2. *Problem* Framing, 3. *Praxis* as Theoretically Informed Practice, 4. *Project*-Centered Pedagogy, 5. Appreciative Inquiry *Principles*, 6. *Process*-Oriented Document Design, 7. Community *Partnerships*, 8. *Public* Imagination, 9. Student-Generated Information *Products*, and 10. Openness to *Possibilities*. I narrate the story of this teaching experience from an inductive perspective to delineate the key takeaway points of these 10 "P's in a Pod" and to highlight the synergistic discoveries along the way.

Part 1: The Power of Positivity

How I came to care about rhinos is a story both of serendipity and synchronicity. It came as news to me when Dr. Diana Deblanc told me during a barn call for my laminitic mustang, Pilar (fig. 11.1), that horses are distant cousins of rhinos. The fact that Pilar was over four hundred pounds overweight at the time and resembled a baby rhino may have been the conversation starter with Dr. Deblanc that snowy winter day that ultimately launched the Rhino Crash project with my ENGL 413 students. Pilar was a Bureau of Land Management mustang rescue from the Jarita Mesa Wild

Figure 11.1. Pilar, recovering from laminitis at the author's home in Edgewood, New Mexico, January 2020.

Horse Territory near the villages of Los Ritos and Pilar in Northern New Mexico. She had been food-deprived as a feral foal and her growth stunted to a sturdy fourteen hands. The first two veterinarians that came to evaluate Pilar's condition were less than optimistic because of her metabolic history. One might say that Pilar has an obsessive-compulsive eating disorder; hence the four-hundred-pound weight gain and laminitis episode that brutal January. How I came to own a blue-roan Spanish mustang with a propensity for binge eating that looks like a baby rhino is yet another story.

Dr. Deblanc shared with me during her house visit that she serves as a resident veterinarian for a rhino rescue project in South Africa. "You do what?" I queried as Dr. Deblanc arranged a series of tiny needles at various ting points in Pilar's feet and legs. I wanted to know on what family tree horses and rhinos could be considered distant cousins. Archeologists claim that rhinos have been roaming this earth for some 50 million years. The existence of rhinos as a living trace of the dinosaurs certainly contributes to their iconic status as a species. Descending from the order of Perissodactyla, rhinos and horses are ungulates and share several common traits. Both species are hind-gut fermenters as well as sporting an odd number of toes. I had never considered horses' hooves as toes but, on closer examination, their hooves are actually oversized toes. And for a horse with laminitis, those oversized toes are indicators of the health of their digestive systems. In brief, a horse binge eating on high-sugar-content hay can result in laminitis (inflammation of the lamina in their hooves) as well as spike equine metabolic syndrome. Pilar was suffering from both conditions.

Dr. Deblanc's first trip to the rhino conservation sanctuary was in 2017 as a volunteer at the Care for Wild (CFW) Rhino Sanctuary in South Africa, the largest black rhino conservation area in the world. After her first three-week volunteer stint, the CFW offered Dr. Deblanc a staff position for the summer of 2019. During her three-month staff appointment at the sanctuary in 2019, she administered veterinarian care for orphaned baby rhinos left to perish beside the mutilated bodies of their mothers in the South African wilderness, the victims of poaching operatives. She also provided veterinary care for the many horses and dogs used as antipoaching patrol support animals.

Although the South African government owns the rhino sanctuary, it does not provide the necessary operational funding to support and maintain the sanctuary. The CFW Rhino Sanctuary relies on donations to fund its mission to protect, rescue, rehabilitate, and release all threatened rhinos. Fund raising organizations, like the Council of Contributors for whom Dr. Deblanc serves as a member of the board of directors, are critical to the mission and vision of the CFW Rhino Sanctuary. I was immediately intrigued by the international network of support required to protect this endangered species in South Africa.

The spiraling concentric circles of attachments and interspecies relationships extending across the globe on behalf of the rhinos were inspiring to me. *Positivity*, as reflected in the rhino conservation efforts of Dr. Deblanc and the Care for Wild Rhino Sanctuary in the face of overwhelming despair, therefore represents the first "P" of technical communication for environmental action. The telos of rhetorical action rests in the capacity to act toward positive ends that emerge out of a positive sense of purpose. Modeling how to engage that power of positivity represents the first and foremost step of environmental education pedagogy. Showing my students how a positive theory of change is essential to an optimal outcome was critical to the success of this project.

Part 2: The Problem

The following day I told my ENGL 413 students about my conversation with Dr. Deblanc and the two other veterinarians I had consulted over the past week. I explained the challenges I was facing trying to help Pilar recover from her bout of laminitis. We dialogued openly about the divergent perspectives of the different veterinarians and the dismal prognosis for Pilar's condition. I then posed the question: What if Pilar's veterinarian and I applied a positive theory of change to her treatment program? Although the odds were not in Pilar's favor, we all agreed taking a positive frame to the problem was the preferable course of action. Hence, I entrusted Pilar to Dr. Deblanc's care, integrating traditional medicine including acupuncture and chiropractic treatment with contemporary veterinarian methods. We then invited Dr. Deblanc to deliver a guest lecture about her rhino conservation work to our class.

We hosted the event on March 10, 2020, at the University of New Mexico (UNM) Student Union two weeks before the campus closed for the remainder of the semester due to COVID-19. My students and I found ourselves navigating all new territory. None of us knew anything about rhinos before the lecture. Dr. Deblanc packed the room with students and community members for a 60-minute presentation about her experiences rescuing and treating baby rhinos. We learned about the ecological and economic issues threatening the future existence of rhinos: the market demand for rhino horn stems from the historical misrepresentation of its mythic medicinal properties. In brief, rhino horn is believed to have magical aphrodisiac qualities, especially for men. Rhino horn is an ancient cultural symbol of power, virility, and fertility. Profit-driven poaching syndicates work to satisfy market demand even at the highest levels of government. Dr. Deblanc explained that crime networks support rhino-poaching syndicates with payoffs to diplomats who enable smuggling through sealed diplomatic bags to transport rhino horns. "Rhino horn is the highest paid commodity on the black market," noted Dr. Deblanc. "It is far more valuable than gold or blood diamonds."

A kilogram of rhino horn shavings can command prices upward of $16,000.00 sold in black markets in places like Nhi Khe, a small village on the outskirts of Hanoi, Vietnam. Nhi Khe is the global trade capital known for dealing in endangered species. Rhino horn is still openly traded to tourists and Vietnamese consumers despite international restrictions against the buying and selling of rhino products. The shocking claims and experiences delivered in Dr. Deblanc's presentation aligned with reports by other rhino conservationists. Millions of dollars of illegal contraband flow from Africa to Asia in marketplaces through vendors trading in rhino horn, ivory, pangolin, and tiger products (Fowlds & Spence, 2019). According to the International Union for Conservation of Nature's (IUCN) statement concerning black rhinos, conservation efforts are having a positive impact. The IUCN's March 2020 statement reports that between 2012 and 2018, the black rhino (*Diceros bicornis*) population across Africa had grown at a modest annual rate of 2.5 percent from an estimated 4,845 to 5,630 animals in the wild.[1]

Even with this hopeful trend, it became clear to us that the poaching crisis is far from over with approximately 2.4 African rhinos poached every day, or one every 10 hours. The horizon to extinction for the black rhino is only five years if nothing changes. The death rate in 2020 surmounts birthrates of black rhinos tenfold. As Dr. Deblanc asserts: "How arrogant are we as humans to think it is OK for any species to go extinct on our watch. Every species matters—pangolins, rhinos, whales, elephants. It is not our right to remove these from existence. We need to stand up and educate and help in any way we can so that future generations can appreciate these creatures and not look at them as a relic from the past."[2] The call to action moved us. I watched the students' faces, identifying with their reactions of horror and grief throughout Dr. Deblanc's presentation. She gave witness to the personal risks to safety and threats of harm, and even death, that conservationists and orphanage staff members face every day. Dr. Deblanc's sense of optimism and commitment represented an important motivator for my class. "With the mission to help every rhino, rather than focusing on the losses, I have to focus on the efforts of keeping the survivors healthy for generations to come."[3] Dr. Deblanc's journey invited us into the collective suffering of South Africa's interspecies communities.

My students and I were able to meet in-person one more time before COVID-19 shelter-in-place orders closed the campus. We spent that last full class session processing our experience together, as students shared their own sense of shock and horror at what Dr. Deblanc had shown us during her lecture. Personally, I was unable to sleep for weeks as I tried to process the images of dismembered rhinos and their orphaned infants seeking nourishment and safety beside their mothers' carcasses. We remembered Dr. Deblanc's disclosure that local community members near the rhino sanctuary often participate in rhino poaching out of economic necessity, revealing how victims and perpetrators of poaching are caught together in the same violent cycle. "If it is a choice between feeding your families or slaughtering a wild animal, what is the most reasonable thing for a person to do?" she asked the audience during her presentation.

The complexities of the economics and ethics of rhino poaching had to inform our approach to the project. We imagined together how the Rhino Crash project might weave the ethical strands of environmental justice on behalf of human and nonhuman species into a process of generating technical communication products on behalf of the threatened rhinos. *Problem* framing as the second "P" of this pedagogical heuristic for teaching technical communication for environmental action exemplifies the necessity of engaging the complexities and contradictions of environmental issues. Problems like biodiversity depletion, endangered species, and extinction threats are dynamic and multifaceted issues that demand ethical, epistemic, and educational engagement from all available fields of study. Even more, the capacity to identify with the collective suffering of humans and nonhumans as co-inhabitants bound together in our current ecological crisis is critical to effective environmental education. Biophilia, the love of life in all its forms shown in figure 11.2, is the quintessential

Figure 11.2. Diana Deblanc and baby black rhino Zwazi at the Care for Wild Rhino Sanctuary, South Africa, 2019.

response to our collective suffering. How we frame issues like interspecies suffering as a problem is therefore essential to realizing productive outcomes in teaching technical communication for environmental action.

Part 3: Praxis—Engaging Theory through Practice

Dr. Deblanc's presentation cultivated deep identification within my class for the threatened rhinos. Put more simply, we fell in love with rhinos. Dr. Deblanc's personal experience had, in a very real way, put her on a first-name basis with the rhinos. Her direct contact with the rhinos, in turn, infused what would evolve into the "Rhino Crash" project with the necessary sense of presence. Engaged relationships like the one Dr. Deblanc embodied with the black rhinos offered us the occasion to be present with her, the rhinos, and the issues they were facing together. Moved by these intricately braided attachments, we worked together for the next eight weeks to generate a set of educational information products for Dr. Deblanc's fund-raising campaign to support the rescue efforts in South Africa. We wanted to generate a set of infographics that cultivated a strong sense of identification within a broad demographic of stakeholders.

Many cutesy, kitschy rhino-themed products abound in the public sphere such as "Save the Chubby Unicorns" t-shirts, coffee mugs, and handbags. We wanted to avoid the cliché and design more educationally generative information products for diverse users and ecological citizens. Making the distinction between audience as citizen rather than simply audience as consumer was a critical decision for us. Moreover, we did not want to reduce the suffering and violence to mass marketing appeal for general consumption. My students and I also decided together that exploiting the shock effect—the horror and aversion we had felt at the violent images of rhino poaching—would not only risk alienating audiences but could also prevent them from becoming concerned global citizens invested in sustained engagement with the issue.

We needed to find a textual space for environmental rhetoric between "cutesy kitsch" and "shock and horror." My students and I also considered the ethical dimensions of the project as technical communicators. The burden of the task was suddenly real. How do we ethically represent the issue of poaching and the threatened condition of the black rhino in South Africa?

Environmental citizens do not simply consume information. Citizens generate, circulate, and metabolize information. We imagined a vibrant ecology of information products that Dr. Deblanc and other advocates could engage within their different circles of belonging: classrooms, faith communities, workplaces, civic organizations, student clubs, and so on.

With these goals in mind, we looked to the indigenous ethical principles of *ubuntu*, as described by Archbishop Desmond Tutu in his reflections about his efforts in the post-apartheid Truth and Reconciliation hearings, to inform the rhetorical aims of our project. Tutu reflects: "Our nation sought to rehabilitate and affirm the dignity and personhood of those who for so long had been silenced, had been turned into anonymous, marginalized ones. . . . Ultimately this third way was consistent with a central feature of the African Weltanschauung—what we know in our language as *ubuntu*. . . . It is to say, 'My humanity is caught up, is inextricably bound up, in yours.' We belong in a bundle of life" (1999, p. 31). Recognizing the dignity, intrinsic value, and inalienable right of rhinos to exist as a species represents an ethically consistent position from an *ubuntu* worldview. Tutu's depiction of *ubuntu* as an ecology of braided attachments helped my students and me situate rhinos "in a bundle of life" through an interstitial ecology of texts.

Within the South African Sotho language system, *ubuntu* is not only an ethic. It is a theory of human psychology as well as a rhetorical paradigm. *Ubuntu* provides the topoi, places for finding and making arguments. *Ubuntu* as an indigenous ethic reaffirms a worldview of interspecies interdependence. Moreover, the narrative experiences offered by Dr. Deblanc gained greater efficacy for us in light of an understanding of *ubuntu*. As Cana Uluak Itchuaqiyaq argues in chapter 1 of this volume, "Narratives are a method of communicating important local expertise that is often overlooked as 'scientific' knowledge in attempts to understand complex environmental situations." Inviting my students to apply the notion of *ubuntu* alongside our reading of Robin Wall Kimmerer's article "Weaving Traditional Ecological Knowledge into Biological Education" helped students engage environmental action from a more holistic vision that aligned traditional indigenous ecological wisdom with Western systems of knowledge making. Traditional ecological knowledge, as Kimmerer argues, is reflected in indigenous cultures all over the world. It represents the "long intimacy and attentiveness to a homeland and can arise wherever people are materially and spiritually integrated with their landscape" (Kimmerer, 2002, p. 433).

Praxis, the third "P" in teaching technical communication for environmental action, engages theory with practice, moving us toward a more holistic vision. In terms of the Rhino Crash project, our class sought not only to align our vision with the indigenous South African ethic of *ubuntu* as praxis to resist the commodification of nature and culture, but more broadly to cultivate our attunement with nature through technical communication integrating traditional ecological knowledge with Western epistemologies.

Part 4: The Project

My students and I agreed that the necessity of hope would be essential to move our audience as citizens toward environmental action. We moved forward inspired by the concept of *ubuntu* (the recognition that we were all bound up in this broken world together) and with confidence in the conviction that we could engage together in the devastating struggle with joy and hope. The students themselves came up with the name "Rhino Crash" for our team effort. It was a tongue-in-cheek rhetorical move on their part. Whereas a constellation of lions is commonly termed a "pride," a gathering of wolves is known as a "pack," and a group of elephants is called a "herd," a community of rhinos is termed a "crash" to denote the power and impact of their singular gravitas and collective force. Similarly, from the point of view of my students as emerging technical communicators, the sudden and cataclysmic shock of the pandemic had impacted them in ways they could only describe as "crashing down their lives" as we worked remotely together toward the production of educational materials on behalf of rhino conservation. The metaphor of "crash" (as a verb and a noun) that signifies both an incident of ruination and a symbol denoting a constellation of rhinos aptly captured the semiotic tension of the project.

We imagined the Rhino Crash project as a digital archive documenting product development at each step of the process as well as creating an open-source platform for a suite of infographics for local and global citizens. With the construction of my ENGL 413 teaching website, we were able to circulate work-in-progress drafts (using password-protected student folders), publish the production process of the project, establish an archive of documents, and provide an open source of public information products for diverse users. Through a network of teleconferences, email correspondence, and the teaching website platform, we were able to mitigate the constraints of the shelter-in-place quarantine and complete the Rhino Crash project as envisioned. One of the most compelling outcomes of this class was the growing ecology of relationships that formed around the Rhino Crash project, a wide circle of beneficiaries directly impacted by our collective efforts. From the students themselves as stakeholders to

Figure 11.3. Black rhinos at the Care for Wild Rhinos Sanctuary, South Africa, 2019.

Dr. Deblanc as our partner, extending outward to the many advocates, leaders, and caretakers working to protect the rhinos, the Rhino Crash project ultimately benefited the rhinos like those in figure 11.3 as our target outcome.

Part 5: The Principles—Appreciative Inquiry

Building a scaffolding of generative principles for document design was critical to the success of the Rhino Crash project. Applying an Appreciative Inquiry model in my ENGL 413 technical communication course for the Rhino Crash education project helped us achieve a number of course learning outcomes as well as project goals. The notion of *ubuntu* aligned synergistically with an Appreciative Inquiry model for technical communication product development.[4] Appreciative Inquiry was first implemented with a partner organization, Save the Children foundation in Zimbabwe, through the 1990s to provide access to clean water to local communities. "One aspect that differentiates Appreciative Inquiry from other visioning or planning methodologies is that images of the future emerge out of grounded examples from an organization's positive past," explains David

Cooperrider et al. (2000, p. 13). The confluences between the goals for our ENGL 413 technical communication class and an Appreciative Inquiry model were significant.

The "4D Cycle" as introduced in *Appreciative Inquiry: Rethinking Human Organization toward a Positive Theory of Change*, delineates a recursive process of invention. It organizes action around these heuristics: Discovery ("What gives life?"), Dream ("What might be?"), Design ("What should be?"), and Destiny ("How to empower?"). The fifth "D," Define ("What is?"), was later included in the Appreciative Inquiry model, expanding the framework into what is now termed "the 5D cycle."

Appreciate Inquiry first reinforces a positive approach to problem framing that draws productively on the past but remains open to transforming the future. Second, Appreciative Inquiry offers an accessible heuristic for invention toward empirical research that can effectively engage novice technical communicators in the recursive process of inquiry. Third, Appreciative Inquiry is grounded in principles of imaginative intellectual play and moves researchers toward positive outcomes in the active process of vision making. Finally, Appreciative Inquiry helps vision makers and document designers resist despair, fatalism, and self-fulfilling narratives of structured failure.

By engaging the Rhino Crash project through stories of possibility, positive action, and environmental relationship, my students and I sought to enact the work of technical communication as acts of ecocitizenship. Creating technical communication documents through the lens of Appreciative Inquiry resulted in a generative inductive process, ultimately infusing the students' final products with a sense of joy, opportunity, authority, whimsy, and playfulness and inviting users into a story of hope rather than despair.

Appreciative inquiry helped my students discover, discern, and articulate what the needs, problems, and action items that advocates like Dr. Diana Deblanc and the frontline workers at the Care for Wild Rhino Sanctuary were asking global citizens and allies to fulfill. As such, Appreciative Inquiry promoted a dialogic process of needs assessment between my students and Dr. Deblanc in her role as a subject area expert. Through a sustained process of discovery and dialogue, we learned how the braided attachments connecting us to this vulnerable and distant species and the community working to protect it could actually make a difference.

Confronting deficiency narratives and resisting the despair of fatalistic thinking represented some of the greatest challenges we faced. Appreciative Inquiry helped us to counter these grand narratives by constructing a collective vision toward a positive revolution of change. Using Appreciative Inquiry *Principles* as the fifth "P" of this environmental pedagogical model helped to ground our project ethically and empirically. Using these as our first principles along with the ethics of *ubuntu* promoted a recursive process to teaching technical communication for environmental action.

Part 6: The Process

My students and I spent the first eight weeks of the Spring 2020 semester reading, writing, and discussing the assigned readings for my ENGL 413 Science, Medical, and Environmental Writing class.[5] We also engaged in a series of field exercises as part of the inquiry process, which included an environmental field site visit, a public rhetoric analysis exercise, and a subject area expert interview. By the time Dr. Deblanc delivered her lecture on March 10, we had already completed the first eight weeks of the syllabus and my students had finished all field exercises and foundational assignments. When pandemic shelter-in-place orders closed the campus 10 days later on March 20, we were ready to launch into their capstone team project proposals.

Each team submitted a group project proposal, working drafts of information products, transmittal memos with project assessments, and final drafts for feedback from Dr. Deblanc and myself. These team working documents were then uploaded to a password-protected platform to establish a digital archive for the Rhino Crash project. Upon completion of the project, we published the Rhino Crash digital archive of team-generated information products on an open-source platform for free access by Dr. Deblanc, the Council of Contributors, educators, and other advocates as users.

This process-centered approach to technical communication pedagogy required a high tolerance for ambiguity within the classroom. The key discovery from a process approach revealed that teaching technical communication for environmental action ultimately engages the disturbance ecology of local and global conditions at both the macro and micro-levels and indexes the collective suffering of a planet in crisis. Life gets messy. All of us in the class learned to exercise a high degree of flexibility and a willingness to engage the ever-changing rhetorical situation influencing project development. *Process* as the sixth "P" of this teaching modality ultimately decenters the classroom, invites questioning, and redistributes authority to the students themselves as they navigate the ambiguities and challenges of a threatened generation of human and nonhuman species.

Part 7: The Partnerships

The partnerships formed between Dr. Deblanc and the students ultimately created a team of technical communicators and ecological advocates. As the subject area expert for the rhino conservation efforts in South Africa, Dr. Deblanc was central to the evolution of the project. She served as the community liaison between us, the Care for Wild Rhino Sanctuary (fig. 11.4), and the fund-raising nonprofit organization Council of Contributors (CoC). This triangulation constituted the sphere of influence within which we were able to exercise our efforts and expertise as technical communicators from within our own environmental sphere of concern. The needs and outcomes articulated by Dr. Deblanc and her appeals for help in sustaining her efforts on behalf of this international rhino conservation coalition provided the necessary exigence for classroom-community partnership.

Figure 11.4. Dr. Deblanc bottle feeding orphaned baby rhino Zwazi.

That exigence, however, needed the spark of energy and imagination of student leaders within the classroom to move that vision into action. That *chispa* of imagination and energy was ENGL 413 student Isabel Strawn (fig. 11.5). In her capacity as the Lobo Gardens Club president, Ms. Strawn directly facilitated and promoted Dr. Deblanc's lecture at the UNM Student Union, as well as serving in a class working group that eventually constructed the Rhino Crash project website.[6] Isabel reflects on her experience: "It made me feel powerless and made me want to figure out how to help. I believe the other students also felt something similar. The lecture was a catalyst to creative, out of the box solutions because of how aggressive it was."[7] *Partnerships* as the seventh "P" for teaching technical communication for environmental action reinforces the necessity of cultivating networks of information and support that extend beyond the academic institution to help students learn to write for public and

Figure 11.5. Isabel Strawn, UNM Lobo Gardens president, January 2020.

professional audiences. Moreover, it helps us as teachers and students to demystify the specialist discourses of the academic inquiry for greater access and usability. The most profound takeaway of this seventh "P" of environmental pedagogy is the gift of guides and mentors for helping us navigate the environmental turbulence and ecological crises facing our planet. The network of collaborative relationships of the Rhino Crash project poignantly reflects the larger ecologies within which human and nonhuman species exist. Solutions to the greatest challenges rest in synergistic efforts and deep connections between humans and nonhuman species (fig. 11.6).

Part 8: The Public Imagination

Situating student-generated information products for the Rhino Crash project within the public sphere represented the central challenge of our endeavor. We quickly found ourselves navigating a semiotic universe of

Figure 11.6. Dr. Deblanc and baby rhino Zwazi enjoying the sun at Care for Wild Rhino Sanctuary, 2019.

complex images and contradictory tropes. My students found themselves straddling a number of complex tasks as they constructed audiences for their information products. Who has the power to enact change? Who is actually capable of saving the rhinos?

We wanted to resist reconstructing a narrative of hopelessness and despair. On the other hand, we didn't want to minimize the severity of the threats to rhinos and their advocates by generating a set of products around feel-good Pollyanna tropes for easy public consumption. Even questions about which type of rhinos to represent in our information product campaign were raised. South African rhinos with two horns (white and black rhinos) are much larger than other types of rhinos, such as the Sumatran and Javen rhinos that have only one horn. All rhinos are critically endangered with some of these species having less than six hundred animals left in the wild.

The fetishization of the rhino in the public imagination is fueled by a market economy around rhino products—both in the black market and in the free market. Consequently, my students and I quickly found ourselves in a rhetorical hall of mirrors. The circulation of rhino conservation rhetoric through a mesh of intertwined egocentric, ethnocentric, anthropocentric, and ecocentric discourses sustains attention and energy through some measure of fetishization. How does the rhino as a metaphor accumulate social value within complex cultural rhetorical ecologies of the public sphere? In what ways does the public imagination perpetuate the topos of rhino as fetish (both a symbol of power, desire, money, and sex as well as an iconic symbol of threatened potentiality, nature, ontological substance, and beingness)?

Because of the always-and-already quality of environmental rhetoric as a type of fetishization of nature, engaging a purity of aims within an ecocentric discourse appears to be impossible. In *Green Culture: Environmental Rhetoric in Contemporary America*, Carl Herndl and Stuart Brown (1996) offer a rhetorical model for environmental rhetoric that foregrounds the aims of regulatory (ethnocentric rhetoric), scientific (anthropocentric rhetoric), and poetic discourses (ecocentric rhetoric) to make visible their various rhetorical aims. Close examination of the public rhetoric of the rhino reveals that economic discourse (an egocentric rhetoric) represents another discursive formation that significantly conditions the rhetorical situation within which technical communicators must shape information products.

For example, a preliminary key word search for the term "rhino" during the early stages of this research immediately revealed a variety of rhino-related products available in the free market. An array of merchandise such as t-shirts and coffee mugs with monikers invoking the childlike "save the chubby unicorns" motto appeared among the search findings. Masking the violence and bloodshed, rhinos as "chubby unicorns" exist in a whimsical market universe alongside search findings trading on the myth of rhino horn's fertility and virility properties.

The rhino, like other endangered species and threatened environmental conditions, exists in a regulatory, scientific, economic, and cultural rhetorical tug of war. As a trope, my students and I struggled to construct our educational information products within the cacophony of the public sphere that resisted perpetuating the fetishization (and exoticization) of the rhino. Is the rhino as emblazoned on t-shirts, facemasks, and handbags any less a fetish than the rhino sold as aphrodisiac on the black market? Is it possible to cultivate identification and imagination in environmental rhetoric without invoking desire and the fetishization of nature?

My provisional answer to that question is that environmental rhetoric unavoidably fetishizes nature through regulatory, scientific, economic, and poetic discourse due to our desire to know it, protect it, and contain it rhetorically. As such, the eighth "P" in teaching environmental communication must include the *Public* sphere as the ideological ecosystem perpetually generating the topoi that set the limits of our rhetorical imagination. Moreover, the technical communication classroom does not exist in a rhetorical vacuum. Much of the difficult work of the technical communicator is navigating the discourses of the public sphere that shape our assumptions and arguments implicitly and explicitly. Tropes of nature in the public sphere are saturated with values (intended and unintended meanings). Environmental rhetoric is a semiotic mesh from which we cannot extricate ourselves.

Part 9: The Products

How did my students resolve this environmental conundrum of discursive aims within the complex cultural rhetorical ecologies of the public sphere? The students organized into six working groups to develop rhino information products over the final eight weeks of the semester.[8] Each individual group member generated a project proposal using an Appreciative Inquiry invention heuristic. Coalescing their individual proposal into a team vision statement, the students then derived and designed sets of rhino information products for diverse target audiences.

Team one produced a set of products for a fund-raising campaign. Using a series of social media platforms such as Twitter and Instagram, team one created information graphics depicting a rhino mother and baby using hashtags and Instagram messages. The evocative images and warm palette of color choices delighted the class. However, the first sketch inadvertently represented the one-horn rhino Asian species, not the two-horn South African black rhino. None of us knew the difference. With the guidance of Dr. Deblanc, team one revised their representation of the South African rhino.

The promotional concept for "Buy a Cup of Coffee for a Rhino" campaign ultimately evolved into a fund-raising proposal to invite college students as the target audience to donate the cost of a cup of coffee to the Council of Contributors to support rescue efforts at Care for Wild Rhinos Sanctuary in South Africa. The enthymematic message of team one's social media campaign was succinctly summed up in three words: "Buy or Save." The message adroitly extends the opportunity for the audience to engage as environmental citizens, not just consumers.

Team two imagined and generated a grant proposal to support a public art project. The team researched and proposed the construction of an art installation of a hornless rhino sculpture to promote public education about endangered species. They drafted a companion brochure representing the rhino art installation describing their vision: "The sculpture stands 6 x 12 feet of gray steel and depicts a white rhino with large horn jaggedly cut from the bottom. The skin of the rhino is etched with local tribal symbols such as waves and spirals to give a local touch to global issues. The size of the rhino depicts both the realistic size of the creature as well as the impact of removing it from the environment." Team two's detailed description of their imaginary sculpture invokes *ubuntu* environmental ethics, as well as representing the violence of poaching.

Team three developed a presentation applying Appreciative Inquiry principles to the Rhino Crash project as a whole, seeking to provide an organizing vision and ethical confluence to synergistically guide the six different project teams. The dynamic Prezi flow chart poignantly exemplifies the alignment of traditional ecological knowledge with science toward a restorative vision for rhino conservation.

Team four generated a suite of posters, flyers, and social media products targeting college students across the disciplines to stir environmental awareness about the plight of endangered species like the rhino. In dialogue with Dr. Deblanc, team four foregrounded the issues of threatened species for college-level students across the disciplines. The aim was to reach beyond STEM disciplines and engage students in other fields as well in the work of environmental conservation.

Team five designed the Rhino Crash website with a vision toward establishing and promoting a new UNM Rhino Conservation Club. Isabel Strawn served as the team leader and web designer. She describes the design choices and rhetorical features of their project:

> A main feature we added to the site to invoke pathos were pictures of baby rhinos. Any baby species elicits an involuntary dimorphous expression. The use of yellow helped us evoke hope. We were careful not to allow the design of the page to be distracting and focused on infantilization of the rhinos as seen in the book project thumbnail. Pictures of baby rhinos with their mothers as well as with Dr. Deblanc allowed us to insinuate that rhinos need caretakers like mothers to their children. This is the ultimate connection between humans. We all have/ had mothers or something of the same nature.[9]

The Rhino Crash website was designed to serve as the public platform for the future student UNM Rhino Conservation Club.

Finally, team six developed a constellation of products to promote future leadership in environmental conservation. One brochure directly applied the recommendations of Dr. Deblanc's lecture advocating for young women to professionalize in veterinary medicine and international conservation. Team six effectively framed the imagination of prospective audiences by illustrating the possibilities of pursuing careers in environmental conservation as researchers, practitioners, antipoaching security forces, teachers, and activists.

Products, the ninth "P" of this environmental communication pedagogy, as opportunities for invention and engagement, helped to cultivate a classroom culture that promoted both critical thinking and imagination. Above all, promoting the development of culturally appropriate, ethically informed information products designed to teach, delight, and move audiences to right action in relation to both culture and nature (fig. 11.7), represents the principle goal of technical communication for environmental action.

Figure 11.7. Sample "Buy a Cup-of-Coffee for a Rhino" campaign poster.

Part 10: The Possibilities

Rhino Crash as an environmental pedagogical project represented an occasion for my students to develop a deep and sustained identification with a difficult and deeply complex environmental issue. In practice, it became an opportunity for transforming our shared anxiety in the midst of the pandemic toward collective positive social action. My students individually and together not only achieved the learning outcomes for the course under the constraints and disruptions of the shelter-in-place mandates, but they also initiated an environmental action project with both local and international impacts. Above all, the Rhino Crash project gave my students the occasion to exercise agency and engage in advocacy work toward something bigger than themselves—the larger ecosystem within which we are all interdependent. As a threatened generation of humans organizing efforts on behalf of an endangered species of nonhumans, my students had the occasion to situate themselves within the larger environmental picture through language.

The Rhino Crash project helped to make visible our abiding kinship ties across species. It exemplified the work of environmental stewardship by foregrounding the ethical imperatives of traditional ecological knowledge and indigenous cultural practices like *ubuntu* that honor the rights of nature and cross-species interdependence. Teaching technical communication for environmental action, therefore, must engage language itself as habitat. How we inhabit and use language ultimately constructs the environment about which we write and semiotically represent. Navigating the ethno-centric (utilitarian), anthropocentric (scientific), ecocentric (poetic/arts), and egocentric (capitalist) discourses conditioning our relationships to nature represents the challenging terrain technical communicators must traverse. Exemplifying the aims at work in technical communication is more than a pedagogical imperative. It is our ethical duty.

The key question for us as teachers of technical communication is: Which environmental discourses are we asking students to reproduce? Reflecting on her experience sponsoring and working on the Rhino Crash project, Isabel Strawn observes:

> It made me realize the actual distress we currently live in. It's not distress for just the planet but all its creatures as well. During the lecture we learned about the cost of one rhino horn. The poaching of the rhinos is not just a result of cruel and unwarranted violence. It is a response to self-preservation. The same type that can be seen throughout most species on the planet including plants. Working on the project made me realize that there are no monsters, just us. There is no specific bad guy that must be stopped. We are all the bad guy.
>
> I think the biggest challenge with engagement is that environmental issues must compete with the perfectly packaged products that offer short-lived, instant gratification. Addressing environmental issues calls for a disruption of the mundane as it envelops your life. It means working hard for no guarantee of anything in return. Working to save the planet consists of being self-less. Living in a world that does not reward a giving nature but takes advantage of it is the biggest challenge in not only getting millennials to care about the environment but any human being that is part of this Earth's society.[10]

The exigence for the rhino conservation project, woven into the strands of interspecies attachments and more-than-just-human connections, ultimately served to support the success of the class and the Rhino Crash education project. We learned that rhinos as an endangered species index a larger ecosystem in distress wherein threatened human and nonhuman species together face habitat deterioration, natural resource depletion, poaching, predation, displacement, and mutual extinction. We are "bundles of life" all bound up together.

Conclusion

Teaching technical communication for environmental action represents the most urgent charge and most significant pedagogical challenge of our professional lives. One of the most compelling discoveries for my students and me within the complex chain of ecological relationships of the rhino project was the recognition of the deeply intertwined interspecies interdependence and the challenges of representing these connections in technical information products. As my students and I realized together, current ecological interstitial relationships reveal that human actions on the environment do matter, positively and negatively. We empirically observed the ways in which the rhino's future, like the fate of my horse Pilar, was inextricably tied to human agents. Likewise, we learned how the future existence of endangered species like rhinos is intricately interwoven to human actors, directly and indirectly.

Taking an Appreciative Inquiry perspective to an environmental problem to enact principled praxis shaped by the processes of building project-focused partnerships helped my students generate information products that represented these complex ecological relationships. Engaging the aforementioned "P's in a Pod" for teaching technical communication cultivated the conditions for a number of other possibilities. First and foremost, this conceptual framework allowed students to foreground their own imagination and appreciation as they engaged complex environmental issues. It created a pedagogical space open to the serendipity and synchronicity of discovery. Additionally, it illustrated the value of foregrounding an ethics-centered approach to audience analysis and document design.

Resisting market-driven models of consumerist audience action approaches, my students learned to engage audiences as global citizens and environmental advocates. Moreover, they generated information products with a long usability shelf life. Over a year later, Dr. Deblanc continues implementing the fund-raising campaign designed by my ENGL 413 students and finding new applications for the materials my students designed. In August 2020, Dr. Deblanc raised over $6,700.00 during the annual World Rhino Day to support antipoaching efforts in South Africa—generating the largest pool of donations since she launched her rhino awareness efforts in New Mexico.

If we apply 10 "P's" for teaching technical communication for environmental action, we can strategically weave the basic tenets of ecoliteracy into the fabric of our courses. We can illustrate to our students the reciprocal nature of all our relationships in and beyond the classroom, reminding us that not only does the earth care for and sustain us as humans and nonhumans alike, but we must, in turn, take care to sustain the earth. As a final case in point, Pilar's laminitis offered us a starting point for applying the 10 "P's in a Pod" in an intentional and systematic way. The dismal prognosis offered by the first two veterinarians who evaluated Pilar before Dr. Deblanc's consult in January 2020 indicated that euthanasia was a likely, final outcome of her deteriorating condition. The choice to seek a different outcome informed by a positive theory of change was our first step. The arduous caretaking required to restore our planet and all our relations (humans and nonhumans) is our collective task.

When we teach about environmental challenges like poaching and biodiversity depletion, we need to help our students consider not only what these losses mean to us as humans but what they mean for other species and the larger ecosystem. Transformation is the ultimate possibility. The loss of the rhino represents something far more profound than the extinction of the iconic "chubby unicorn." It indexes global system failure. We can seek a different outcome.

Isabel Strawn and Krishna Patel are two of my ENGL 413 students who wanted to seek a different outcome beyond our Spring 2020 technical communication classroom. Together with Dr. Deblanc and other community members and educators, Ms. Stawn and Mr. Patel helped to establish a nonprofit organization at the space between the academic and the public spheres to sustain the Rhino Crash project through the Southwest Environmental Education Cooperative (SWEEC) in the fall of 2020.[11] By invoking the "P's in a Pod" heuristic of invention, we are currently expanding this community-engaged project. We are refining and reproducing the pedagogical practices of our ENGL 413 Science, Medical, and Environmental Writing classroom in public spaces such as community gardens, libraries, parks, and open spaces with community writing workshops and immersive environmental education installations. The SWEEC team plans to launch the Rhino-Pa-Looza Immersive Mobile

Environmental Education Exhibit featuring what we call a "low tech and low brow" art installation of a homemade rhino sculpture named "Riley" (sourced and welded from materials recycled found in wrecking yards of vintage automobile parts). We will be towing Riley across the state of New Mexico and the Southwest region to bioparks and zoos on a flatbed truck in the summer of 2022 for an environmental education and fund-raising campaign on behalf of the rhinos.

In closing, I am also happy to report our binge-eating mustang remains sound and fully recovered over a year later, completely restored to her ever-so-feisty and slightly svelte self.

Figure 11.8. Pilar and her riding companion Anna Michelle Broom, back in the saddle again, June 2020.

Notes

1. International Union for Conservation of Nature's (IUCN) statement of Black Rhino Endangered Species, March 19, 2020, available at https://www.iucn.org/news/species/202003/conservation-efforts-bring-cautious-hope-african-rhinos-iucn-red-list.

2. Diana Deblanc, "Our Precious Few," invited lecture, University of New Mexico, Albuquerque, March 10, 2020, https://unmecoliteracy.wordpress.com/2020/03/02/upcoming-event-our-precious-few-with-dr-diana-deblanc/.

3. Diana Deblanc, phone interview with the author, October 24, 2020.

4. Current developments in Appreciative Inquiry have expanded the process of invention to the "5D Cycle" adding "Define" to the heuristic. See Appreciative Inquiry: https://appreciativeinquiry.champlain.edu/learn/appreciative-inquiry-introduction/5-d-cycle-appreciative-inquiry/.

5. ENGL 413 Science, Medical, and Environmental Writing Course website available at https://www.michelle-kells.org/engl-413.

6. Isabel Strawn and Rhino Conservation Team 5 ENGL 413 Rhino Crash project website at https://rhinocrashcon.wixsite.com/unmrcc.

7. Isabel Strawn, phone interview with the author, November 1, 2020.

8. ENGL 413 Rhino Education Projects Teams 1–6 available at https://www.michelle-kells.org/rhino-conservation-projects.

9. Isabel Strawn, phone interview with the author, November 1, 2020.

10. Phone conversation between the author and Isabel Strawn, May 12, 2020.

11. See Southwest Environmental Education Cooperative https://www.southwestenvironmentaleducationcooperative.org/.

References

Berry, W. (2015). *It all turns on affection*. Counterpoint.

Cooperrider, D. L., et al. (Eds.). (2000). *Appreciative inquiry: Rethinking human organization toward a positive theory of change*. Stipes.

Fowlds, G., & Spence, G. (2019). *Saving the last rhinos: The life of a frontline conservationist*. Little, Brown.

Herndl, C. G., & Brown, S. C. (Eds.). (1996). *Green culture: Environmental rhetoric in contemporary America*. Madison, WI: University of Wisconsin Press.

Kimmerer, R. W. (2002). Weaving traditional ecological knowledge into biological education: A call to action. *BioScience, 52*(5), 432–438.

Tutu, D. (1999). *No future without forgiveness*. Doubleday.

Epilogue: Right Relation with the Whole World

Creating a Richer Polyvocality for Environmental Technical Communication

CAROLINE GOTTSCHALK DRUSCHKE

I am writing, in some ways, from outside of technical communication, or at least from a complex intersection of technical communication, rhetoric of science, environmental communication, community-engaged research, and straight up environmental science. But maybe that's useful to the reader. Like some of you, my academic graduate training didn't come through technical communication, but I found myself drawn to this scholarship out of necessity. In the first decade of the 2000s, as I worked to connect—or maybe more accurately to reconcile—my critical theory–heavy training as a graduate student in rhetorical studies with my teaching and administrative work in community-based writing, and an increasing amount of coursework, collaboration, and public engagement related to the ecology of human-altered landscapes, I searched for colleagues who found some sense in that intersection.

For me, that initial "aha!" moment came from Stuart Blythe, Jeffrey T. Grabill, and Kirk Riley's "Action Research and Wicked Environmental Problems: Exploring Appropriate Roles for Researchers in Professional Communication." Published in 2008 in the *Journal of Business and Technical Communication*, that article, as many readers will know, describes the authors' community-centered work related to a public engagement process on a proposed dredging project that would have had profound environmental impacts on that community. Right there in the abstract, I stumbled on the approach I had been searching for. In no uncertain terms, Blythe, Grabill, and Riley linked environmental technical communication with action research, insisting on putting community concern first. As they declared: "The authors argue that the primary goal of action research related to environmental risk should be to identify and support the strategies used by community members rather than to educate the public" (272).

And their epigraph! A quote from painter dian marino in her book *Wild Garden: Art, Education, and the Culture of Resistance* that emphasized narrative, community, and mundane struggle: "Personal stories and social histories of resistance and change, the failures no less than the successes, need to be widely shared. Otherwise, we are left with the impression that community issues and struggles are born out of nothing—or that only extraordinary heroic people can get involved and make a difference" (30–31). This article was my introduction to environmental technical communication, though admittedly a very specific approach. This emphasis—on engaging closely with environmental issues, learning about them from multiple angles, engaging with communities to "identify and support the strategies used by community members" rather than (so-called) solutions identified by (so-called) university experts, attending to questions of social justice, collaborating with ordinary folks who become heroes, telling stories of change—put many things in right order for me. It gave me some comfort that there might be a home for this kind of work, and I'm delighted to see that focus continue to grow with this edited collection. Crucially, I am not suggesting here that there is only one way into this work, only one genealogy. Simply a kinship. A holding together of like concern.

For me, that concern has increasingly centered on work that connects community-driven approaches to the joys and challenges of living with streams and rivers. This place-based research has focused on watershed-based agricultural conservation in eastern Iowa (Druschke, 2013, 2018) and community approaches to fish passage and dam removal in New England (Druschke et al., 2017; Druschke and Rai, 2018; Druschke, 2019). Since relocating to Wisconsin in 2017, it has focused squarely on community responses to increasingly frequent and severe flooding exacerbated by climate change in the underresourced southwestern corner of the state. That work involves a sometimes overwhelming amount of cross-disciplinary knowledge, as I work to understand the violent history of Euro-American settlement in the region and how it reshaped stream systems and landscapes, geomorphology and hydrology of southwestern Wisconsin streams, contemporary changes in precipitation and how that shapes flood peaks, narrative approaches to community storytelling about streams and floods, tools necessary to communicate in compelling ways across genres related to community input and advocacy, policies that shape flooding and flood resilience in the region. That work also involves a not insignificant amount of emotional labor and trauma, supporting community members to tell their stories about catastrophic floods, while supporting undergraduate and graduate students through their own trauma as they support these efforts. It also involves time: time to get 100 miles away from campus to spend in the community, time spent at public meetings and community events, time listening and sometimes getting yelled at, time learning and building relationships that last beyond academic semesters and run up against tenure clocks or the contingency of many academic positions.

These are challenges that I see addressed straight on in the chapters gathered in this volume: the need for cross-disciplinary knowledge, the weight of relationship building and trauma, the frequent incompatibility of community-focused work with university structures. But in case after case in this collection, the authors offer readers equipment for engaging these challenges and working through them to support community-driven environmental action with an eye toward equity and justice. They share like concern.

The task of writing the epilogue to this collection, then, was intimidating but also exhilarating. Early drafts of its essays have already begun to shed relief on and shape my own work, and I am confident their published form will do the same for you. Even my title—"Right Relation with the Whole World: Creating a Richer Polyvocality for Environmental Technical Communication"—was directly inspired by the work gathered here.

"Whole world" is a phrase borrowed from Cana Uluak Itchuaqiyaq's chapter, "When the Sound Is Frozen: Extracting Climate Data from Inuit Narratives," a phrase expanded on from Itchuaqiyaq's own father, Caleb Lumen Pungowiyi, a Yupik leader, climate change researcher, and advocate for Indigenous context-based knowledges. As Itchuaqiyaq explains in their chapter: "TC as a field has historically leaned toward 'scientifically measurable' data in its research, but as Pungowiyi argues, this focus is too narrow when it comes to issues like climate change. If Arctic researchers focus only on scientifically measurable data and continue to ignore the potential contextual data contained in local user narratives, then they are missing the opportunity to see 'the whole world.'" "Whole world," then, includes "scientifically measurable data" alongside "the potential contextual data contained in local user narratives." And it's a perspective, I would argue, that threads together all the chapters in the collection. The authors gathered here—from Josephine Walwema's focus on Cape Town water policy in "Participatory Policy: Enacting Technical Communication for a Shared Water Future," to Sara Parks and Lee Tesdell's focus on agricultural water quality in "Resilient Farmland: The Role of Technical Communicators," to Daniel P. Richards's focus on flood inundation and insurance in "Flood Insurance Rate Maps as Communicative Sites of Pragmatic Environmental Action," to Barbara George's focus on fracking risks in "Health in the Shale Fields: Technical Communication and Environmental Health Risks"—attend to "the science," but they work, too, to engage with context and narrative and to put all those different ways of understanding "whole world" in conversation with one another. That synthetic work to create and honor larger, more complex conversations, these chapters collectively argue, is essential to technical communication for environmental action. Addressing whole world, as we see in this collection, demands the recognition, incorporation, and amplification of many human and nonhuman actors and many ways of knowing the issue at hand. Intervening, ethically, in environmental issues demands this whole world perspective.

Directly to this point, the methodological and stylistic innovation of Lauren E. Cagle and Roberta Burnes's coauthored chapter, "Collaborating for Clean Air: Virtue Ethics and the Cultivation of Transformational Service-Learning Partnerships," offers what I take to be essential terminology for apprehending whole world, alternating between their voices to create what they describe as "a richer polyvocality." That polyvocality, we learn through their chapter, is built through the labor of reciprocal relation: through "humility, patience, respect, honesty, and curiosity." I propose that Cagle and Burnes could easily be describing not just their collaboration, but this collection as a whole, and doing so in two distinct ways:

- as a collection that, taken together, represents a polyvocality of approaches to environmental technical communication that forward values like humility, patience, respect, honesty, and curiosity to intervene in environmental concerns, and

- as a group of interventions that, in each individual case, emphasize the necessity of polyvocality in environmental action to apprehend something more like whole world in order to take ethical environmental action.

Crucially, that work is positioned, in chapter after chapter, as dependent on emplacement and attention to wider context: to whole world and to the polyvocality necessary to address, include, account for, and be accountable to whole world. Taken together, the chapters gathered in this collection offer a polyvocal vision for the future of environmental technical communication that emphasizes further polyvocality. These chapters emphasize process, context, and advocacy in environmental action and technical communication to invite readers to join in this shared work. To be clear, this is not an uncritical approach to simply including more people at the table in every conversation related to environmental conflict. Instead, these chapters are attentive to power and, thus, pay careful attention to expanding a conversation about environmental conflict to the folks most affected by it and most excluded from it, finding new platforms from which they might be able to speak. The chapters' generally careful attention to equity and justice—a point to which I will return—make that discernment possible.

I have highlighted some very broad connections across the chapters gathered here, but how else can we begin to pull these threads together, and where might they go? At the outset of this collection, editor Sean Williams offered what I take to be very helpful insight into three key mechanics that hold these chapters together:

- *praxis*, or "practical applications of technical communication theories such as social justice, participatory design, community action, service learning, and ethics,"

- *phronesis*, or "practical wisdom of communities," and

- *dialogue*, or "good-natured collaboration."

I found those markers incredibly useful heuristics for diving into these rich chapters at the outset, and I want to close out this text with a few additional heuristics for understanding what I see as some major contributions and shared points of emphasis for the chapters gathered here:

- innovative process

- scalar connection

- improvised action

- right relation

I then turn to where I think this collective work might direct us or could go. Taken together, I see these chapters as a manifesto for the future of environmental technical communication: offering a roadmap for where the field has been, is now, and might—or even *must*—be headed in response to the increasingly urgent demands of environmental degradation and conflict.

Innovative Process

Many of the chapters in the collection work to critique existing processes and create new ones, often offering models for the reader to follow in their own work or to learn from to create their own. Daniel Card, for instance, in "Boundary Waters: Deliberative Experience Design for Environmental Decision Making," advocates for prospective, not just retrospective work: "the work that must precede effective intervention." Card reimagines a public engagement process for a proposed mine just outside the Boundary Waters Canoe Area in the Upper Midwest US that could foster equitable decision making and procedural justice. In her chapter, "Participatory Policy: Enacting Technical Communication for a Shared Water Future," Josephine Walwema also focuses on public participation and water, using critical discourse analysis to interrogate Cape Town, South Africa's response to a massive water shortage. Her chapter details Cape Town's successful work to adapt their approach by enlisting city residents into the work of adapting to drought. As she describes, "This practice has reshaped the public's relationship to water." Technical communication, in these chapters, becomes a lens through which to not only critique but also to propose alternative modes of public intervention in environmental conflicts.

Moving from public participation into pedagogy, Bob Hyland's "In Defense of Greenspace: Students Discover Agency in the Practice of Community-Engaged Technical Communication" reshapes the technical communication classroom to enroll students in the work of protecting a campus-adjacent greenspace while reenvisioning the temporal constraints of the semester-long course. Building an environmental action campaign that spanned multiple semesters "provide(d) opportunity for our students to see that technical communication must be used to make a difference." Monika A. Smith, in "Writing for Clients, Writing for Change: Proposals, Persuasion, and Problem Solving in the Technical Writing Classroom," shares this emphasis on sustaining relationships and projects across semesters, working long-term with university facilities management staff on campus-based sustainability challenges to upend the common approach to semester-long community-based courses. So do Lauren Cagle and Roberta Burnes, in "Collaborating for Clean Air: Virtue Ethics and the Cultivation of Transformational Service-Learning Partnerships," who attend to long-term pedagogy but emphasize the important process of building relationships across university instructors and community practitioners. Like Hyland and Smith, they work beyond the constraints of the academic semester to build a lasting relationship between the two of them, and that relationship, in turn, supports student learning and intervention in community-identified needs. Through what they call "participatory action teaching," Cagle and Burnes rework the idea of university service learning and remake the process of building sustainable university-community partnerships. Their chapter narrates the very process that brought the two of them together. As they describe: "PAT [participatory action teaching] is expansive; it's not just about teaching within the confines of a college class, but about a messy ongoing relationship of which our coproduced courses are just one manifestation, and in which there is room for so much more: joint workshops and informal exchanges of expertise and former students becoming organizational collaborators and even the coproduction of knowledge by, for example, our coauthorship of the very book chapter you're reading right now." Michelle Hall Kells, in her chapter "Rhino Crash: Teaching Science, Medical, and Environmental Writing for Social Action," also focuses on the classroom, introducing what she calls "the 10 pedagogical 'P's in a Pod'" to introduce a process for enlisting students in environmental action in ways that responded to the needs of the community partner as well as to the students' needs.

Inside and outside the classroom, this collection offers examples of its authors thinking creatively to remake university, community, and social processes to offer richer opportunities for community connection and inspired environmental action. Scholars, teachers, and practitioners of technical communication all have roles to play not only in critiquing unjust processes—though that is an important first step—but also in supporting the creation of new ones. There are increasing chances for doing that work in our classrooms, where we often have some level of autonomy over syllabus design, and many of the collection's authors offer detailed inspiration for how to transform some of the academic structures that get in the way of deeply reciprocal, sustainable university-community relations. That work should continue, especially as it extends beyond individual classrooms to create pedagogical structures that allow graduate students and contingent faculty to participate more easily. And there is much more room, I would argue, for technical communicators to intervene in environmental public processes, finding creative ways to embed themselves in communities and government structures to advocate on behalf of more equitable inclusion for environmental action. This is no doubt challenging work, but technical communicators must work creatively to identify partnerships, grant opportunities, and collaborations to intervene in mundane but centrally important ways in the workings of public engagement processes.

Scalar Connection

Many chapters in this collection explicitly focus on working across scales: intervening at the local level in global issues like climate change, not surprisingly, but also working to connect across campus, city, and regional scales. Beth Shirley's chapter, "The Narrative of Silent Stakeholders: Reframing Local Environmental Communications to Include Global Human Impacts," is maybe the most explicit in theorizing this scalar work, adapting the concept of "societal teleconnections" from the field of sociology to daylight the global ramifications of local actions and insist that technical communicators advocate for affected persons away from the site of debate. As Shirley argues, "Existing models of stakeholder engagement appear to situate local and global environmental concern squarely against local stakeholders' interests (economic or otherwise), when the long-term, global, environmental impacts are also going to affect local stakeholders, if they are not already." The work of technical communicators, then, becomes the work to amplify and advocate for local interests against hegemonic global powers. Dan Richards's "Flood Insurance Rate Maps as Communicative Sites of Pragmatic Environmental Action" similarly focuses on local responses to global climate change, inserting himself into what he describes as mundane work to protect local homes from increased flooding. Richards "asks the reader to consider the more procedural, banal work that can be done still in the meantime as we figure the bigger, more wicked things out." This hyperlocal work, these authors tell us, matters in the big picture and in the small.

We see this argument, too, in several chapters that attend to the technical writing classroom as a space for exploring scalar connections. Monika Smith, for instance, describes building a relationship with campus facilities management to provide opportunities for international students to make local, place-based interventions that help to combat global climate change. As Smith explains, writing for the environment inspired her students to see their campus not only as a specific place but also as part of a global system. For Robert Hyland and Michelle Hall Kells, too, campus-adjacent partnerships offer students opportunities for global interventions related to land conservation, for the former, and species conservation, for the latter. This local scale, we are meant to understand, is where global change happens.

These scalar connections seem centrally important in the current moment—when globally pervasive degradation has the potential to overwhelm—and also more possible—given the increasing opportunities for networked communication. Practically speaking, these societal teleconnections offer a manageable way into seemingly intractable problems. When Monika Smith's computer science students, for instance, design plans to reduce electricity consumption and plastic waste on their own campus, they chip away at seemingly intractable abstractions like global climate change. There is something important to be said for the provision of hope in projects like these. I am writing this epilogue in summer 2021, during a massive global health crisis, on the heels of a deadly heat wave in the US Pacific Northwest, amid rampant wildfires across western North America, in a week during which the Gulf of Mexico literally caught fire. Hope seems in short supply at the moment. But these local, tractable, addressable issues offer us and our students ways to *do something*, as Dan Card suggests. These scalar connections also matter for understanding complexity and nuance: something like whole world. These connections offer ways of considering how actions at the community scale impact and are impacted by the municipality, the watershed, the state, the region, the bioregion. They offer important—and mostly still underexplored—means for decentering the US experience in environmental technical communication. I look forward to future work that builds from the non-US focus of chapters like Josephine Walwema's and Beth Shirley's that invite collaborations between US and non-US technical communication scholars and practitioners to more richly and equitably understand and intervene in the differential impacts of global forces.

Improvised Action

Taken together, the chapters in this collection identify the challenges of trying to respond to and intervene in wily, complex environmental conflicts that are unfolding in real time. The best of this work is timely, responsive, emergent, *kairotic*. But it's also messy and uncomfortable, and it inevitably leads to failures alongside successes. Dan Richards addresses this philosophy head-on as he reflects on his role building an interactive flood insurance calculator, in a short period of time, in a process already determined by others—a situation no doubt familiar to other technical communicators. As Richards reflects, "I didn't consult any technical communication theory, didn't weigh different scholarly approaches. I had to learn as much as I could about flood insurance and coding. I didn't have time for anything else. I just had to build *the thing*." And yet, his training—and all of ours—is always close to hand, even if not to mind. Like Debra Hawhee's (2004) focus on the ancient links between rhetoric and athletics in Ancient Greece, our training is in our bones: embodied preparation that allows us to respond flexibly in the moment to changing conditions. Cagle and Roberta Burnes, too, are explicit about this philosophy of responsivity and flexibility. They emphasize "the kinds of naturally occurring conversations that have led us to a point where we have a collaboration worth theorizing in the first place." Meanwhile, their hesitance to resist the label of "service learning" offered important possibility: "not needing to settle on a label for our work was valuable, because it left us rather unconstrained in imagining what our collaboration might become, beyond a single transactional service-based semester." Based in part on inspiration of Phelps-Hillen (2017), Cagle and Burnes insist, "As we became more comfortable and secure working together, we transitioned from an early transactional approach to this kind of transformational partnership, in which both Cagle and Roberta's goals and identities are not fixed, but rather open to change based on 'emergent possibilities' and increased willingness to move 'beyond the status quo.'" Michelle Hall Kells also focuses on the classroom, considering how this philosophy of improvisation comes into play as students design a range of timely interventions into black rhino conservation, using the tools at hand to reach a range of audiences to support their cause.

Other authors in the collection drill down to consider how improvisation works at the level of community organizing. Sara Parks and Lee Tesdell, for instance, look for creative interventions into the negative environmental impacts of American industrial agriculture. As they detail in their first case study, the Des Moines Water Works case was ultimately a failed attempt at intervention in declining water quality in the state, but it also served "as a first attempt and a model for more lawsuits." In other words, it reset the terms of possibility for fighting water pollution in the state on the policy front, just as their second case, the Prairie Strips initiative, did at the level of the individual landowner, encouraging farmers to look differently at the possibilities for corners of their own farms. Barbara George, too, looks to creative intervention in "Health in the Shale Fields: Technical Communication and Environmental Health Risks," where she explains that a lack of state attention to environmental health risk encountered by residents led to the emergence of alternative community supports. When the state failed to take these concerns seriously, community efforts intervened to center intersectional environmental justice.

Environmental technical communication, by definition, engages with wicked problems, which Dan Card nicely captures in his chapter as "problems that involve multiple interacting systems, uncertainty, and value-laden solutions." As he continues, "Given the sociotechnical nature of wicked problems, equitable solutions do not arise from technical expertise alone. Each possible solution will affect various stakeholders differently, and that makes these problems 'public' in an important sense. In environmental policy contexts, recognizing a problem as wicked requires us to figure out how to bring stakeholders to the table—not only those with expertise in geology, ecology, economics, and so on, but also those impacted by the problems, solutions, and alternatives under consideration." I think we, as a field, need to be straightforward about the fact that we are engaging with environmental challenges that don't have singular "right" answers. There may be "right" answers scientifically for specific problems, "right" answers from the perspective of regulators or landowners or politicians, "right" answers in the eyes of particular stakeholders. But not one singular "right." That's what makes each of these cases so interesting, so important, and so dynamic. These are problems that, by design, demand improvisation. And it is only through continued improvisation—repeated, career-long, collaborative investment in responsivity and adaptation—that we can build up a repertoire that allows us to respond and intervene—*ethically*—in each case.

Right Relation

Last, but certainly not least, I want to draw the reader's attention to the ways that many of the authors gathered here model an urgent concern with what I am calling "right relation," a deep concern with putting themselves in constellation with the human and other-than-human aspects of environmental conflict and holding themselves accountable to those relations.

This, I think, is the most radical and most important axiom that emerges from this collection: that environmental technical communication should concern itself with equity, action, and justice-centered intervention. Not as afterthought or add-on, but as central tenet.

For Barbara George, this is a focus on "intersectional environmental justice." For Michelle Hall Kells, this is "ecological stewardship through interspecies kinship ties." For Monika Smith, it is "caretaking." For Cagle and Roberta Burnes, this is a feminist concern with "virtue ethics," what they describe as "an ethical framework rooted in the cultivation of personal virtues, practical wisdom, and contextualized decision making." Many of these chapters build from the important lessons of Natasha Jones (2016, 2017; Jones et al., 2016; Jones & Walton, 2018), Rebecca Walton (Walton & Jones, 2013; Walton et al., 2019; Walton & Agboka, 2021), Donnie Johnson Sackey (2018, 2020), and others related to social justice and environmental justice in technical communication, and environmental technical communication would do well to continue to move social justice and environmental justice concerns to the absolute center of the field. Likewise, there is a wealth of resources in the *CCCC Black Technical and Professional Communication Position Statement* (Mckoy et al., 2020) that could inform, enlighten, and enliven environmental technical communication, including long traditions of Black user experience design, Black rhetorics of health communication, and Black social movements and community work. As they work through different terminology and different intellectual traditions, these chapters articulate a compass that centers them, their work, and their students' work in relation to and with accountability to others. Environmental technical communication here becomes a toolkit to create a livable future for whole world.

I find Cana Uluak Itchuaqiyaq especially compelling on this point, offering an approach concerned with human and more-than-human relations. As Itchuaqiyaq describes,

> In my Inuit worldview, I rely heavily on a *kin*centric perspective—where human agency is tempered by, and in concert with, the agency of the world *in which* humans participate. In other words, humans are positioned as *a* POV. I do not consider human agency as the central, determining, or strongest vector in the web of interconnectedness of what is known as the world. It isn't the "world around us" at all; rather, it is, like my father states, the "whole world."

I want to insist, as I close out this epilogue, that the real work for environmental technical communication comes from where relationality takes us. I want to insist that the work we're collectively forwarding in this volume is not without heartache and risk. It is not without screwing up and trying to do better and screwing up again. This iterative work—to build relationships over time and space; to interrogate the ethics of our work with communities, students, families, and self; to hold ourselves accountable to nonacademic partners who don't care about our tenure reviews and course schedules; to respond flexibly and sometimes naively and always improvisationally to urgent environmental concerns; to take unpopular positions against more powerful adversaries; to sometimes alienate ourselves from university colleagues and supervisors—is where environmental technical communication must go.

This work is urgent, and it is not without risk. So, I repeat the powerful question that closes Itchuaqiyaq's chapter and that I have been repeating to myself since first reading it: "Tell me: What risk have you incurred?"

In the shared vision offered in this text, the future of environmental technical communication demands putting ourselves in right relation with the world to act, risk and all, on those relations.

References

Blythe, S., Grabill, J. T., & Riley, K. (2008). Action research and wicked environmental problems: Exploring appropriate roles for researchers in professional communication. *Journal of Business and Technical Communication, 22*(3), 272–298.

Druschke, C. G. (2013). Watershed as common-place: Communicating for conservation at the watershed scale. *Environmental Communication: A Journal of Nature and Culture, 7*(1), 80–96.

Druschke, C. G. (2018). Agonistic methodology: A rhetorical case study in agricultural stewardship. In C. Rai & C. G. Druschke (Eds.), *Field rhetoric: Ethnography, ecology, and engagement in the places of persuasion* (pp. 22–42). University of Alabama Press.

Druschke, C. G. (2019). A trophic future for rhetorical ecologies. *Enculturation: A Journal of Rhetoric, Writing, and Culture.* https://www.enculturation.net/a-trophic-future

Druschke, C. G., Lundberg, E., Drapier, L., & Hychka, K. C. (2017). Centering fish agency in coastal dam removal and river restoration. *Water Alternatives, 10*(3), 724.

Druschke, C. G., & Rai, C. (2018). Making worlds with cyborg fish. In B. McGreavy, J. Wells, G. F. McHendry Jr., & S. Senda-Cook (Eds.), *Tracing rhetoric and material life: Ecological approaches* (pp. 197–222). Palgrave.

Hawhee, D. (2004). *Bodily arts: Rhetoric and athletics in Ancient Greece*. University of Texas Press.

Jones, N. N. (2016). The technical communicator as advocate: Integrating a social justice approach in technical communication. *Journal of Technical Writing and Communication, 46*(3), 342–361.

Jones, N. N. (2017). Modified immersive situated service learning: A social justice approach to professional communication pedagogy. *Business and Professional Communication Quarterly, 80*(1), 6–28.

Jones, N. N., Moore, K. R., & Walton, R. (2016). Disrupting the past to disrupt the future: An antenarrative of technical communication. *Technical Communication Quarterly, 25*(4), 211–229.

Jones, N. N., & Walton, R. (2018). Using narratives to foster critical thinking about diversity and social justice. In A. M. Hass & M. F. Eble (Eds.), *Key theoretical frameworks: Teaching technical communication in the twenty-first century* (pp. 241–267). University of Colorado, Utah State University Press.

marino, d. (1997). *Wild garden: Art, education, and the culture of resistance*. Between the Lines.

Mckoy, T., Shelton, C. D., Sackey, D., Jones, N. N., Haywood, C., Wourman, J., & Harper, K. C. (2020). *CCCC Black technical and professional communication position statement with resource guide*. Conference on College Composition and Communication. https://cccc.ncte.org/cccc/black-technical-professional-communication

Phelps-Hillen, J. (2017). Inception to implementation: Feminist community engagement via service-learning. *Reflections: A Journal of Community-Engaged Writing and Rhetoric, 17*(1), 113–132.

Sackey, D. J. (2018). An environmental justice paradigm for technical communication. In A. M. Hass & M. F. Eble (Eds.), *Key theoretical frameworks: Teaching technical communication in the twenty-first century* (pp. 138–160). University of Colorado, Utah State University Press.

Sackey, D. J. (2020). One-size-fits-none: A heuristic for proactive value sensitive environmental design. *Technical Communication Quarterly, 29*(1), 33–48.

Walton, R., & Agboka, G. Y. (Eds.). (2021). *Equipping technical communicators for social justice work: Theories, methodologies, and pedagogies*. University Press of Colorado.

Walton, R., & Jones, N. N. (2013). Navigating increasingly cross-cultural, cross-disciplinary, and cross-organizational contexts to support social justice. *Communication Design Quarterly Review, 1*(4), 31–35.

Walton, R., Moore, K. R., & Jones, N. N. (2019). *Technical communication after the social justice turn: Building coalitions for action*. Routledge.

Contributors

Roberta Burnes is the environmental education specialist for the Kentucky Division for Air Quality (DAQ) in the Energy and Environment Cabinet (EEC). She runs the division's education and outreach program, traveling across the commonwealth to engage students with air quality and its impact on human health. Burnes serves as DAQ's public information officer; she runs the website, writes stories, and develops educational media. Since beginning her partnership with Cagle, Burnes now trains EEC staff on improving digital accessibility of the cabinet's website and documents. She is a certified Kentucky master environmental educator.

Lauren E. Cagle is an assistant professor of writing, rhetoric, and digital studies (WRD) and associate faculty in environmental and sustainability studies (ENS) at the University of Kentucky. Her research focuses on overlaps among digital rhetorics and scientific and technical communication. Cagle frequently works with local and regional environmental and technical practitioners; her current collaborative partners include the Kentucky Division for Air Quality, the Kentucky Geological Survey, the University of Kentucky Recycling Program, and the Arboretum, State Botanical Garden of Kentucky. Cagle's work has appeared in *Technical Communication Quarterly*, the *Journal of Technical Writing and Communication*, *Rhetoric Review*, and *Computers and Composition*.

Daniel Card is an assistant professor in the Department of Writing Studies at the University of Minnesota–Twin Cities. His research draws on a range of approaches in rhetoric and technical communication to explore how we navigate complexity. He is particularly drawn to problems that bring together a variety of publics with diverse expertises and values, and as such much of his work focuses on problems in the realms of science, technology, medicine, and environment. His recent articles have been published in *Review of Communication*, *Journal of Business and Technical Communication*, and *Rhetoric of Health and Medicine*.

Caroline Gottschalk Druschke is a professor in the Department of English at the University of Wisconsin–Madison. Building from training in rhetorical studies, freshwater ecology, and community engagement, Druschke's collaborative, community-driven work questions how people and rivers continue to change each other in the face of climate change–exacerbated flooding. Dr. Druschke coedited *Field Rhetoric: Ethnography, Ecology, and Engagement in the Places of Persuasion*, has published widely across rhetorical studies and the environmental sciences, and has received funding from the National Science Foundation, the Andrew W. Mellon Foundation, the US Environmental Protection Agency, and the National Park Service.

Barbara George is a lecturer in the Writing and Communications Department at Carnegie Mellon University. Barbara researches the intersections of environmental risk and communication, issues of intersectionality within environmental deliberation, and environmental justice within environmental communication. Additionally, Barbara explores the use of narrative in finding new ways to "reimagine" environmental frames, particularly in Rust Belt communities. Barbara has also researched different student retention efforts: best practices for literacies across disciplines, multilingual initiatives, and writing center supports. Barbara's work appears in *Composition Forum*, *Environmental Sociology*, and *Communication Design Quarterly* among others.

Bob Hyland is an associate professor educator of rhetoric and professional writing at the University of Cincinnati. Bob's research role has evolved over the years. Starting out as a technical writer on a research team doing corrosion studies of drinking water distribution system piping, he primarily conducted literature reviews and packaged research results for publication. Currently, his work focuses on student discovery of agency through community-engaged technical communication pedagogy as well as the efficacy of nature education to improve learning outcomes for underserved youth. He is coauthor of an article and several peer-reviewed proceedings on contaminant accumulation in distribution system piping.

Cana Uluak Itchuaqiyaq is a tribal member of the Noorvik Native Community in Northwest Alaska and is an assistant professor of professional and technical writing at Virginia Tech. Her research combines her academic background in both the digital humanities and physical sciences and currently centers on creating accessible online databases of Inuit knowledges and developing natural language processing techniques to extract climate change data from Inuit narratives. She is an author on the upcoming National Climate Assessment 5, Alaska chapter.

Michelle Hall Kells is a professor of rhetoric and writing in the Department of English at the University of New Mexico where she serves as the translingual literacy studies coordinator. Kells teaches courses in science, medical, and environmental writing; ecopoetics; and language equality education. She serves as chair of the National Consortium of Environmental Rhetoric and Writing (affiliate of the Rhetoric Society of America) and the executive director of the Southwest Environmental Education Cooperative. Her contributions have been recognized by the 2018 and 2013 Best of Rhetoric and Composition Independent Journals Outstanding Essay Awards and 2012 Conference of College Composition and Communication Outstanding Book Award. Kells most recent books include *Vicente Ximenes, LBJ's Great Society, and Mexican American Civil Rights Rhetoric* and a coedited volume with Laura Gonzales, *Latina Leadership: Language and Literacy Education across Communities.*

Sara B. Parks is an assistant professor at Stephen F. Austin State University where she serves the English Department as the technical writing coordinator. Parks teaches courses in technical and scientific writing, digital rhetoric, sustainability literature, editing, and publishing. Her research engagement in environmental resilience efforts began through a position in a National Science Foundation grant on biorenewable energy during graduate school and continues to inform her work.

Daniel P. Richards is an associate professor of English at Old Dominion University. His research has covered a variety of areas, including the politics of higher education, disaster and risk communication, and the intersections of posthuman and new materialist theory in the field of rhetoric. Most recently, his work focuses on the empirical study of visual risk communication, particularly through the lens of community engagement, in order to best understand how to communicate sea level rise and flooding to coastal communities.

Beth Shirley is an assistant professor of technical communications and rhetoric at Montana State University. Beth's research has focused on our rhetorical relationships with technology and the environment and using that understanding to frame science communication for specific populations. Her latest projects emphasize engagement with rural communities and local knowledges through interdisciplinary collaboration toward improved science communication. Her work has appeared in *Technical Communication* and *Present Tense*, along with localized outreach publications.

Monika A. Smith is an assistant teaching professor with the Academic and Technical Writing Program at the University of Victoria, British Columbia, Canada, where she has taught engineering and computer science students the art of crafting professional technical communication for many years. A special area of interest includes finding ways to connect technical writing students with campus-based sustainability initiatives. Currently, she is working on a cross-disciplinary study with coauthors Dr. Ilamparithi Chelvan and Sajib Ghosh to determine whether self-review checklists can encourage students enrolled in electrical engineering courses, which typically offer little or no writing support, to monitor, regulate, and revise their project reports to enhance clarity, professionalism, and readability.

Lee S. Tesdell is a professor emeritus at Minnesota State University, Mankato, and he now lives on his farm near Slater, Iowa. Tesdell completed his PhD at Iowa State University in the rhetoric and professional communication program with Drs. Rebecca Burnett and David Wallace. Tesdell taught courses in technical communication at Mankato and now advocates for conservation and renewable energy through writing, speaking, and holding field days on his farm.

Josephine Walwema teaches technical and professional communication at the University of Washington, Seattle. Her research and teaching interests began with the scholarship of teaching and learning and evolved to social, economic, and environmental concerns. That evolution has led her to examine the reach and impact of TPC with respect to equity and access in the lives of individuals in the North and the Global South. She has authored several articles and book chapters on these subjects.

Sean D. Williams is a professor of technical communication and information at the University of Colorado–Colorado Springs. Sean's research has taken many forms over the years, beginning with information architecture in complex web environments to social media in technology start-ups, to user experience design for 3D virtual reality. Most recently, his work focuses on environmental communication where his central focus is understanding how best to communicate science to drive personal conservation behaviors and public policy changes. He is author or coauthor of two books and more than 50 articles, book chapters, and trade publications.

Index

Note: Page numbers in *italics* indicate figures, and
bold page numbers indicate tables in the text.